STOCHASTIC ANALYSIS OF SCALING TIME SERIES
From Turbulence Theory to Applications

Multiscale systems, involving complex interacting processes that occur over a range of temporal and spatial scales, are present in a broad range of disciplines, from financial trading to atmospheric dynamics. Turbulent flows are a classical example of such a complex system. Several methodologies exist to retrieve multiscale information from a given time series obtained from a complex dynamical system; however, each method has its own advantages and limitations.

This book presents the mathematical theory behind the stochastic analysis of scaling time series, including a general historical introduction to the problem of intermittency in turbulence, as well as how to implement this analysis for a range of different applications. Covering a variety of statistical methods, such as Fourier analysis, structure-function analysis, wavelet transforms, and Hilbert-Huang transforms, it provides readers with a more thorough understanding of each technique and when they should be applied. New techniques to analyze stochastic processes, including empirical mode decomposition, autocorrelation function of increments, and detrended analysis, are also explored.

The final part of the book contains a selection of case studies, on the topics of turbulence and ocean sciences, to demonstrate how these statistical methods can be applied in practice. With MATLAB codes available online, this book is of value to students and researchers in Earth sciences, physics, geophysics, and applied mathematics.

FRANÇOIS G. SCHMITT is Research Professor in the Laboratory of Oceanography and Geosciences, at the Centre National de la Recherche Scientifique (CNRS), France. His research interests include turbulence and nonlinear variability in geophysics, marine turbulence, and multifractal analysis and modeling.

YONGXIANG HUANG is Associate Professor in the State Key Laboratory of Marine Environmental Science at Xiamen University, China. He was awarded the 2013 Division Outstanding Young Scientists Award by the European Geosciences Union in Nonlinear Processes in Geosciences.

STOCHASTIC ANALYSIS OF SCALING TIME SERIES

From Turbulence Theory to Applications

FRANÇOIS G. SCHMITT & YONGXIANG HUANG

CAMBRIDGE
UNIVERSITY PRESS

Shaftesbury Road, Cambridge CB2 8EA, United Kingdom

One Liberty Plaza, 20th Floor, New York, NY 10006, USA

477 Williamstown Road, Port Melbourne, VIC 3207, Australia

314–321, 3rd Floor, Plot 3, Splendor Forum, Jasola District Centre, New Delhi – 110025, India

103 Penang Road, #05–06/07, Visioncrest Commercial, Singapore 238467

Cambridge University Press is part of Cambridge University Press & Assessment,
a department of the University of Cambridge.

We share the University's mission to contribute to society through the pursuit of
education, learning and research at the highest international levels of excellence.

www.cambridge.org
Information on this title: www.cambridge.org/9781107067615

© François G. Schmitt and Yongxiang Huang 2016

This publication is in copyright. Subject to statutory exception and to the provisions
of relevant collective licensing agreements, no reproduction of any part may take
place without the written permission of Cambridge University Press & Assessment.

First published 2016

A catalogue record for this publication is available from the British Library

Library of Congress Cataloging-in-Publication data
Schmitt, Francois G.
Stochastic analysis of scaling time series : from turbulence theory to applications / Francois G. Schmitt &
Yongxiang Huang.
pages cm
Includes bibliographical references and index.
ISBN 978-1-107-06761-5 (Hardback : alk. paper)
1. Time-series analysis 2. Stochastic analysis. I. Huang, Yongxiang, 1982– II. Title.
QA280.S36 2015
519.55–dc23   2015033195

ISBN   978-1-107-06761-5   Hardback

Additional resources for this publication at www.cambridge.org/schmitt

Cambridge University Press & Assessment has no responsibility for the persistence
or accuracy of URLs for external or third-party internet websites referred to in this
publication and does not guarantee that any content on such websites is, or will
remain, accurate or appropriate.

Contents

List of figures		*page* vii
Preface		xxiii
1	Introduction: a multiscale and turbulent-like world	1
	1.1 Data from the real world	1
	1.2 Multiscale phenomena	4
	1.3 The Fourier-based methodology and its potential shortcomings	6
	1.4 Conclusion and further remarks	10
2	Homogeneous turbulence and intermittency	12
	2.1 Introduction	12
	2.2 Richardson-Kolmogorov cascade and K41 relation	15
	2.3 Intermittency and Yaglom's cascade model	18
	2.4 Multifractal properties of multiplicative cascades	23
	2.5 Velocity fluctuations and multiscaling properties	32
	2.6 Passive scalar intermittent turbulence	36
	2.7 Multiscaling formulation of general nonstationary scaling time series	37
	2.8 Intermittency and Lagrangian turbulence	39
3	Scaling and intermittent stochastic processes	41
	3.1 Introduction	41
	3.2 Brownian motion, fractional Brownian motion, and Lévy stable walks	42
	3.3 Simulation of continuous multifractal processes	45
4	New methodologies to deal with nonlinear and scaling time series	56
	4.1 Structure function analysis	56
	4.2 Autocorrelation function of increments	63

	4.3	Maximum probability density function scaling of velocity increments	69
	4.4	Detrended fluctuation analysis	74
	4.5	Detrended structure function	77
	4.6	An interpretation in a time-trequency analysis frame	82
	4.7	Wavelet-based methodologies	83
	4.8	Empirical mode decomposition and Hilbert spectral analysis	89
	4.9	General remarks on the scaling analysis methodologies	99
5	Applications: case studies in turbulence		101
	5.1	Homogeneous turbulence	101
	5.2	Passive scalar	123
	5.3	The Lagrangian Turbulence	135
	5.4	Rayleigh-Bénard turbulent convection	144
	5.5	Two-dimensional turbulence	152
	5.6	General remarks on scaling behavior in turbulent flows	160
6	Applications: case studies in ocean and atmospheric sciences		162
	6.1	Coastal marine turbulence	163
	6.2	Water-level dynamics	165
	6.3	Water quality automatic monitoring	170
	6.4	Atmospheric wind velocity	175
	6.5	Wind power time series	179

References 185
Index 202

Figures

1.1 a) Collected oxygen saturation index on the time period January 1 to December 31, 2010, with a sampling time of 20 minutes. The data were collected by a sensor belonging to the MAREL network (Automatic monitoring network for littoral environment, Ifremer, France). Due to several reasons, e.g., failure of the sensor, there are several missing data points. A strong event was also observed around September 1, 2010, showing the nonstationarity of the process. b) An enlargement part on the time period January 1 to January 10, 2010. c) The time interval δt between two successful measurements. The intermittent distribution of δt demonstrates the problem of missing data or irregular time step. *page* 2

1.2 a) Illustration of the Lagrangian drifter trajectories obtained from the global drifter program. b) Measured pdf of the drifter life-time T. The measured pdf $p(T)$ has an exponential distribution. The drifter collected several physical quantities of the ocean with a sampling time of six hours. This provided a large variation of the data length. 3

1.3 Collected water temperature data on the period March 24 to December 13, 2010. These data are automatically recorded every 20 minutes by the MAREL network. A strong annual cycle is observed. Note that this deterministic-like forcing is a typical structure in the geophysical data, which might induce difficulties in the scaling analysis. 4

1.4 a) A one second portion of the measured longitudinal turbulent velocity in a wind tunnel experiment at Johns Hopkins University with a Taylor microscale based Reynolds number $Re_\lambda \simeq 720$. The measured velocity is fluctuating over a large range of time scales. b) The experimental Fourier power spectrum in a log-log plot. Power-law behavior $E(f) \sim f^{-\beta}$ is observed with a β close to the Kolmogorov 1941 theory prediction 5/3. 5

1.5 a) An example of turbulent velocity along a Lagrangian trajectory obtained from a high-resolution direct numerical simulation with a Taylor microscale-based Reynolds number $Re_\lambda \simeq 400$. A vortex trapping event is observed around $t/\tau_\eta = 100$. b) The measured energy dissipation rate $\epsilon(t)/\langle\epsilon\rangle$ along

	the same Lagrangian trajectory, showing a strong intermittent/nonstationary event around time $50 < t/\tau_\eta < 120$.	6
1.6	a) Numerical solution of the Duffing equation with $\epsilon = 1$, $b = 0.1$, $\Omega = 2\pi/25$ and initial value $[x(0), x'(0)] = 1$. Note that the solution $x(t)$ is smooth, with a mean period $T = 9.524$, corresponding to a domain frequency 0.105 Hz. Hence a frequency $f \gg 0.105$ is unphysical. b) The corresponding Fourier power spectrum. The forcing scale is observed at 0.04 Hz as the first peak in $E(f)$. The second peak is at 0.105 Hz for the domain frequency. High-order Fourier harmonic components are visible for $f \gg 0.105$ Hz. They are required by the linear approximation of the Fourier transform.	8
1.7	a) Reconstructed $\tilde{x}(t)$ from the domain frequency $f_D = 0.105$ Hz. b) Reconstructed $\tilde{x}(t)$ from the first 30 Fourier modes. c) Residue $\delta x(t) = x(t) - \tilde{x}(t)$, in which the thick solid line is for the domain frequency and the thin line is for the first 30 Fourier modes. Note that the high-order Fourier harmonic components are the result of the difference between the Duffing solution $x(t)$ and the sine wave profile.	9
2.1	Illustration of the Richardson-Kolmogorov cascade from large to small scales.	16
2.2	(Color online) Fourier power spectrum of velocity and temperature time series recorded in the atmosphere, displaying 5/3 power law decay over about two decades.	17
2.3	A modern example of a proxy of the dissipation field recorded in the atmosphere. The series is normalized by its mean, showing the large fluctuations.	18
2.4	A schematic representation of a discrete multiplicative cascade.	21
2.5	Mechanism generating long-range power-law correlations: two points separated by a distance r have a common ancestor at step p such that $2^p \approx r$.	23
2.6	Schematic graphical representation of the codimension function, which is increasing and convex. Here the plot is shown for the lognormal case, with different values of μ.	25
2.7	Representation for moments from 0 to 10 of five classical models for the velocity structure functions scaling exponents: the K41 model; the β model with $C_1 = 0.125$ (parameter chosen to be tangent to experimental data at $q = 3$); the lognormal model with $\mu = 0.25$; the log-Poisson model with $c = 2$ and $\beta = 2/3$; and the log-stable model with $\alpha = 1.5$ and $C_1 = 0.15$. The three log-ID models are very close one to each other, until the moment of about 6. Larger moments are very difficult to accurately estimate and need a huge number of data points.	35
2.8	Typical $\zeta(q)$ curve obtained for the velocity and passive scalar turbulence, compared with the Kolmogorov-Obukhov-Corrsin nonintermittent curve $q/3$. For low-order moments the two curves are close and for large moments the passive scalars are much more intermittent. Adapted from Schmitt et al. (1996).	38
2.9	Illustration of the ramp-cliff phenomenon for the turbulent temperature.	38

List of figures

3.1 Simulations of various scaling processes (Brownian motion, fractional Brownian motion, Lévy stable motion).   45

3.2 Typical moment functions $\zeta(q)$ for several cases, Brownian motion, fractional Brownian motion for $H = 0.3$, Lévy stable motion with $\alpha = 1.5$ and a multifractal time series with a nonlinear moment function.   46

3.3 Cone $A_\ell(t)$ chosen for the stochastic integration in the $\{(t, \ell), \ell > 0\}$ plane: (a) the causal case, and (b) the symmetric noncausal case.   49

3.4 Intersection between two cones A_ℓ and $A_\ell(t + \tau)$. The darker zone corresponds to the correlation between the two points $X(t)$ and $X(t + \tau)$; it decreases as a power law and generates the long-range correlation of such process.   49

3.5 Examples of subordinated simulations of a multifractal process, following Equation (3.22), with a lognormal multiplicative cascade with $\mu = 0.2$, and values of $h = 0.3$ (above) and $h = 0.5$ (below).   51

3.6 Different values of $H = \zeta_X(1)$ obtained in a $h - \mu$ plane. The region (1) corresponds to a region with unknown value for H. The nonlinear curve which is at the bottom left corresponds to the condition $K(1/2h) < 1/2h - 1$.   53

3.7 Examples of moving average simulations of a multifractal process, following Equation (3.27) with a lognormal multiplicative cascade with $\mu = 0.2$, and values of $h = 0.3$ (above), and $h = 0.5$ (below).   54

3.8 The plane $h - \mu$ giving the value of the scaling exponent $\zeta_X(q)$ for a lognormal process, according to Abry et al. (2009) and Perpete (2013). For the zone which is left in blank, there is no result for the moment, the convergence is not yet demonstrated. Increasing diagonal: $\zeta_X(q) = \frac{q}{2} - K(q)$; open dots: $\zeta_X(q) = \frac{\mu+1}{2}q - K(q)$ and decreasing diagonals $\zeta_X(q) = qh - K(q)$.   54

4.1 Illustration of the turbulent velocity increment $\Delta_\tau u(t) = u(t) - u(t + \tau)$ obtained from a wind tunnel experiment with $Re_\lambda \simeq 720$ and a time increment $\tau = 0.025$ sec; a) A portion of 0.5 second measured longitudinal velocity $u(t)$; b) A time $\tau = 0.025$ second shift $u(t + \tau)$; c) The measured velocity increment $\Delta_\tau u(t)$. In the context of turbulence, the separation time scale τ is interpreted in the inertial range as the typical time scale of a vortex with time scale τ.   57

4.2 Illustration of weight-function $\mathcal{W}(\ell, k)$ for two different separation scales ℓ and 2ℓ. Note that except for the case $f_\ell = n/\ell$ with $n = 0, 1, 2, \cdots$ all Fourier components have contributions to $S_2(\ell)$.   59

4.3 a) Measured cumulative function $\mathcal{P}(\beta, k)$ for $\beta = 1.4, 5/3, 2,$ and 3. The separation scale considered here is $\ell = 1$. b) The measured large-scale influence $\mathcal{P}_1(\beta) = \mathcal{P}(\beta, 1)$. Note that $\mathcal{P}_1(\beta)$ increases with β.   60

4.4 a) Measured second-order structure-function without (open symbols) and with perturbation (solid symbols) for the same β as in Figure 4.3. b) The influence function $\mathcal{I}(\beta, \ell)$ in dB. Note that the larger the β value, the more $S_2(\ell)$ or $\mathcal{I}(\beta, \ell)$ are influenced by the perturbation.   61

4.5 Contour plot of the measured influence function $\mathcal{I}(\beta, \ell)$ in dB.   61

4.6	Measured structure-function $S_q(\ell)$ for the fBm process with a Hurst number $H = 1/3$ and a data length 10^6 points. Power-law behavior is observed for all q considered here. The inset shows the measured scaling exponent $\zeta(q)$, in which the theoretical value qH is illustrated by a solid line. The errorbar is a 95% fitting confidence limit.	63
4.7	Measured autocorrelation function $\Gamma_\ell(\tau)$ of velocity increment with various separation scale ℓ for a turbulent velocity. The turbulent velocity is obtained from a high Reynolds number wind tunnel experiment with $Re_\lambda = 720$. Inset shows the measured location $\tau_0(\ell)$ for the minimum value of $\Gamma_\ell(\tau)$, in which the solid line indicates the value $\tau_0(\tau) = \ell$. The collapse of the symbols and solid line verifies Equation (4.16b) on a large range of scales.	64
4.8	Measured rescaled autocorrelation function $\Gamma_\ell(\tau)$ of velocity increments with various separation scale ℓ. The collapse of symbols confirm the universal rescaling Equation (4.22).	66
4.9	Measured $\Gamma_o(\ell)$ for fBm with $H = 1/3$ and turbulent velocity with $Re_\lambda = 720$. Specifically for the turbulent velocity, a power-law behavior with a scaling exponent 0.80 ± 0.02 is observed on the range $5 \times 10^{-4} < \ell < 5 \times 10^{-2}$ sec, corresponding to a frequency range $200 < f < 2000$ Hz. The error bar is a standard deviation from 120 realizations. The inset shows the compensated curve to emphasize the observed power-law behavior. For comparison, the compensated curve for the second-order structure function is also shown as a solid line. Note that the plateau for the structure function is much shorter than the one for $\Gamma_o(\ell)$. This phenomenon has been interpreted as the large-scale influence.	68
4.10	Cumulative function $\mathcal{P}(f, 1)$ (dashed line) for the second-order structure function and $\mathcal{Q}(f, 1)$ for autocorrelation function of increments (solid line). Note that $\mathcal{Q}(0.496, 1) \simeq 0$, indicating that the influence of the large scale structure might be canceled by each other for $\Gamma_o(\ell)$.	69
4.11	Experimental probability density function of turbulent velocity increments. The turbulent velocity is obtained from a high Reynolds wind tunnel with $Re_\lambda = 720$. The separation τ is in data points. Note that the location of the maximum value of $p(x)$ is close to zero.	73
4.12	Measured $p_{\max}(\tau)$ for fBm process with a Hurst number $H = 0.5$ and 100 realizations and $L = 10^6$ data points each. The errorbar indicates the standard deviation from these 100 realizations. The measured scaling exponent is $\alpha = 0.50 \pm 0.01$. The inset shows the finite sample size effect obtained from 100 realizations with different data length L. Note that with the increasing of the data length L, the measured α is approaching the given value of $\alpha = 0.5$.	73
4.13	An illustration of the DFA procedure. a) Measured velocity $u(t)$ with a portion of length 0.25 sec. The experimental data are obtained from a high Reynolds number wind tunnel experiment (Kang et al., 2003). b) The measured cumulative function $Y(t)$, in which the thin line is the first-order trend with 0.025 sec. c) The first-order detrended data $Y(t) - P^1(t)$, showing the detail after the detrending.	75

List of figures

4.14 Measured $\mathcal{F}_q(n)$ for a fBm process with a Hurst number $H = 1/3$, in which the solid line is a power-law fit. The inset shows the measured scaling exponent $h(q)$, in which the solid line is the theoretical value $h(q) = (1 + H)q$. The errorbar indicates the 95% fitting confidence limit. ... 76

4.15 Illustration of the first-order detrending analysis for the fBm process with $H = 1/3$ and data length 10^4 points. a) The original time series (grey line) and the linear trend with $\tau = 1000$ data points. b) The detrended time series x_τ. Ideally, the large-scale fluctuation $r \geq \tau$ would then be expected to be removed. ... 79

4.16 Comparison of the Fourier power spectrum for original time series $x(t)$ and the detrended time series x_τ for various detrending scales τ, in which the vertical solid lines indicate the scale $f = 1/\tau$. Visually the large-scale fluctuations $f < 1/\tau$ are constrained, verifying Equation (4.62). ... 80

4.17 Experimental $\mathcal{D}_2(\tau)$ for the fBm process with a Hurst number $H = 1/3$. Power-law behavior is observed as predicted by Equation (4.63). The inset shows the compensated curve with a theoretic scaling exponent 2/3. ... 80

4.18 Experimental $\mathcal{B}_q(\tau)$ for the fBm process with a Hurst number $H = 1/3$ and a data length 10^6 data point: a) Equation (4.65) and b) Equation (4.67). Power-law behavior is observed as predicted by Equation (4.66). ... 81

4.19 Experimental scaling exponent $\zeta(q)$ provided by formula (4.65) (denoted as F1, ○) and (4.67) (denoted as F2, □). The theoretical value $q/3$ is illustrated by a solid line. ... 81

4.20 Illustration of the weight function $\mathcal{W}(\tau, f)$ for structure function analysis (dashed line), detrending analysis (thin solid line), and detrended structure-function analysis (thick solid line). The separation scale $f_\tau = 1/\tau$ is indicated by a vertical solid line. ... 83

4.21 Illustration of a wavelet transform with different scales. It presents a multi-scale decomposition of a given data series. ... 84

4.22 a) Illustration of a Meyer wavelet. b) The measured Fourier power spectrum. It shows a compact support property not only in the physical domain, but also in the spectral domain. ... 85

4.23 Illustration of the Haar wavelet; see Equation (4.74) for details. Note that the Haar wavelet can be linked with the classical structure function via the operation of the increment. ... 86

4.24 Measured qth-order moments $Z_q(\ell)$ for the fBm process with a Hurst number $H = 1/3$ computed using the Haar wavelet. Power-law behavior is observed. The inset shows the measured scaling exponent $m(q)$, in which the solid line indicates a theoretical prediction $q/3$. ... 87

4.25 Measured qth-order moments for the fBm process with a Hurst number $H = 1/3$: a) Wavelet Transform Modulus Maximum $\mathcal{Z}_q(\ell)$. b) Wavelet leader $\mathbb{Z}_q(\ell)$; and c) the corresponding scaling exponents $\zeta_W(q)$ (□) and $\xi(q)$ (○). The theoretical value is illustrated by a solid line. ... 89

4.26 Illustration of the characteristic scale for different methodologies: a) Fourier transform; b) wavelet transform; and c) Hilbert-Huang transform. Note that only the Hilbert-based method allows both frequency- and amplitude-modulation. 90

4.27 A typical Intrinsic Mode Function extracted from experimental data. The local extrema points are denoted by ○. The upper and lower envelopes are denoted by a dashed line. 91

4.28 IMF modes from the nonlinear Duffing (Equation [1.5]). The first IMF mode with a mean frequency ~ 0.1 Hz captures the main variation of the Duffing equation. The IMF $C_2(t)$ corresponds to the external force. The IMF mode C_3 and the residual might be the boundary effect of the EMD algorithm. 96

4.29 Measured instantaneous frequency $\omega(t)$ provided by the Hilbert spectral analysis for the first two IMF modes of the Duffing equation. The dashed line are for 0.105 Hz for the mean frequency of Duffing equation, and 0.04 Hz for the external force, respectively. 96

4.30 Measured instantaneous frequency $\omega(t)$ of the $\omega_1(t)$. The mean frequency is $\langle \omega(t) \rangle \simeq 0.21$ Hz. The dashed line indicates a mean frequency of $\langle \omega_1(t) \rangle \simeq 0.105$ Hz. The measured $\omega_1(t)$ possesses a complete oscillation within one period $\simeq 10$. This is the so-called intrawave frequency modulation, representing a nonlinear distortion. 97

4.31 Extracted IMF modes $C_i(t)$ from a Lagrangian trajectory with a vortex trapping event. Visually, a vortex trapping event with a typical time scale $5\tau_\eta$ is isolated around $t/\tau_\eta = 100$. 98

4.32 Extracted IMF modes $C_i(t)$ from a Lagrangian trajectory with a vortex trapping event. Visually, the vortex trapping event with a typical time scale $\sim 5\tau_\eta$ is isolated around $t/\tau_\eta = 100$. Note that the rest of the vortex trapping event has been set as NaN. 98

4.33 Experimental $\mathcal{L}_q(\omega)$ and $\mathcal{M}_q(\omega)$ for the fBm process with a Hurst number $H = 1/3$. Power-law behavior is observed for all q considered here. The inset shows the measured scaling exponent $\xi(q)$ (○) and $\zeta_H(q)$ (□), in which the solid line indicates the theoretical prediction. 99

5.1 Experimental Fourier power spectrum $E(f)$ for the longitudinal (solid line) and the transverse (dashed line) velocities, in which the Kolmogorov's $-5/3$-law is shown as a dash-dotted line. Power-law behavior is observed on nearly a two-decade range $10 < f < 1000$ Hz with the scaling exponents $\beta = 1.63 \pm 0.01$ and 1.61 ± 0.01, respectively, for the longitudinal and transverse velocity components. The inset shows the compensated curves by using the fitted scaling exponents to emphasize the observed power-law behavior. 102

5.2 Experimental second-order structure functions $S_2(\tau)$ for the longitudinal (○) and transverse (□) velocity components. Power-law behavior is observed on the range $0.0005 < \tau < 0.005$ second, respectively, with scaling exponents $\zeta(2) = 0.67 \pm 0.01$ and $\zeta(2) = 0.65 \pm 0.02$. The power-law range corresponds to a frequency range of $200 < f < 2000$ Hz. The inset shows the compensated curves to emphasize the observed power-law behavior. For display clarity, these curves have been vertical shifted. 103

5.3	Experimental cumulative function $\mathcal{P}(\tau,f)$ for various separation time scales in the inertial range: a) $\tau = 0.005$ (resp. $f_o = 200$), b) $\tau = 0.01$ (resp. $f_o = 100$), c) $\tau = 0.02$ (resp. $f_o = 50$), and d) $\tau = 0.04$ (resp. $f_o = 25$). The frequency has been rescaled as $f\tau$. For comparison, $\mathcal{P}(\tau,f)$ for a Kolmogorov power-law is illustrated as a solid line.	104
5.4	Experimental structure function $S_q(\ell)$: a) $S_q(\ell)$ for the longitudinal velocity; b) compensated curves using the fitted scaling exponent; c) $S_q(\ell)$ for the transverse velocity; and d) the corresponding compensated curves. A nearly one decade power-law behavior is observed on the range $0.0005 < \tau < 0.005$ for both velocity components.	105
5.5	Measured scaling exponents $\zeta(q)$ of the longitudinal ($\bigcirc$) and transverse ($\square$) velocity components on the power-law range $0.0005 < \tau < 0.005$ for $0 \le q \le 6$. The inset shows the extended-self-similarity scaling exponents. The Kolmogorov nonintermittent value $q/3$ is indicated by a dashed line.	106
5.6	Measured singularity spectrum $f(\alpha)$ for the longitudinal ($\bigcirc$) and transverse ($\square$) velocity components. For comparison, the curve for the lognormal formula with an intermittency parameter $\mu = 0.25$ is illustrated by a solid line. The insets shows the measured $f(\alpha)$ provided by the extended-self-similarity method. For the same range of q, the transverse velocity has a wider α and $f(\alpha)$ than the longitudinal one. It has been interpreted as coming from the fact that the transverse velocity is more intermittent than the longitudinal one.	107
5.7	Measured location $\tau_o(\tau)$ for the minimum value of the autocorrelation function of velocity increments: the longitudinal ($\bigcirc$) and transverse ($\square$) velocity components, in which the inertial range $0.001 < \tau < 0.1$ second predicted by the Fourier spectrum is shown by a vertical solid line. The solid line indicates $\tau_o(\tau) = \tau$.	108
5.8	Measured $-\Gamma_o(\tau)$ for the longitudinal ($\bigcirc$) and transverse ($\square$) velocity components. The errorbar is a standard deviation from 120 realizations. Power-law behavior is observed on the range $0.0005 < \tau < 0.02$, with a scaling exponent 0.80 ± 0.02 for both velocity components. To emphasize the observed power-law behavior, the compensated curves using the fitted scaling exponent is shown as an inset. For comparison, the compensated curves for the second-order structure functions are also shown as thick (resp. longitudinal) and thin (resp. transverse) solid lines.	108
5.9	Measured $p_{\max}(\tau)$ and $p_z(\tau)$ for the longitudinal ($\bigcirc$, and $\triangle$) and transverse ($\square$, and $\triangledown$) velocity components. The errorbar is a standard deviation from 120 realizations. Power-law behavior is observed on the range $0.0005 < \tau < 0.01$ sec. The measured scaling exponents of $p_{\max}(\tau)$ are 0.36 ± 0.01 for both velocity components, and of $p_z(\tau)$ are $\alpha = 0.36 \pm 0.01$ and $\alpha = 0.33 \pm 0.01$ respectively for the longitudinal and transverse velocity components. To emphasize the observed power-law behavior, the compensated curves using the fitted scaling exponent is shown as inset. For comparison, the compensated curves with a scaling exponent $\zeta(1) = 0.35 \pm 0.01$ for the first-order structure functions	

are also shown as thick (resp. longitudinal) and thin (resp. transverse) solid lines. 109

5.10 Measured $\mathcal{D}_2(\tau)$ for the longitudinal (O) and transverse (□) velocity components. The errorbar is a standard deviation from 120 realizations. Power-law behavior is observed on the range $0.002 < \tau < 0.05$ sec with a measured scaling exponent 0.68 ± 0.02 and 0.65 ± 0.02 respectively for the longitudinal and transverse components. The inset shows the compensated curve using the fitted scaling exponent. 111

5.11 Measured detrended structure functions $\mathcal{B}_q(\tau)$ on the range $0 \le q \le 4$: a) the longitudinal velocity, and b) the transverse velocity. Power-law behavior is observed on the range $0.002 < \tau < 0.02$ sec. 111

5.12 a) Experimental scaling exponents $\zeta(q)$ is extracted on the range $0.002 < \tau < 0.02$ seconds. b) The corresponding singularity spectrum $f(\alpha)$. For comparison, a lognormal formula with an intermittency parameter $\mu = 0.30$ is also shown as a dashed line. 112

5.13 Experimental first-order detrended fluctuation functions: a) the longitudinal velocity, and b) the transverse velocity. Power-law behavior is observed on the range $0.002 < \tau < 0.02$ seconds. 113

5.14 a) Experimental scaling exponents $\zeta(q)$ extracted on the range $0.002 < \tau < 0.02$ seconds, in which the solid line is the nonintermittent value $4q/3$. b) The corresponding singularity spectrum $f(\alpha)$. For comparison, a lognormal formula with an intermittency parameter $\mu = 0.30$ is also shown as a dashed line. 113

5.15 Experimental wavelet-based qth-order moments ($q = 2, 4, 6$): a) wavelet transform modulus maximum of the longitudinal velocity, b) wavelet leaders of the longitudinal velocity, c) wavelet transform modulus maximum of the transverse velocity, and d) wavelet leaders of the transverse velocity. Power-law behavior is observed on the range $0.0005 < \tau < 0.02$ seconds, corresponding to a frequency range $2000 < f < 50$ Hz. 114

5.16 a) Experimental wavelet transform modulus maximum based scaling exponents $\zeta_W(q)$, in which the solid line is the nonintermittent value $q/3$, and b) the corresponding singularity spectrum $f(\alpha)$. c) Experimental wavelet leaders based scaling exponents $\xi(q)$, and d) the corresponding singularity spectrum $f(\alpha)$. For comparison, a lognormal formula with an intermittency parameter $\mu = 0.30$ is also shown as a dashed line. 115

5.17 Experimental probability density function of the number of extracted intrinsic mode function of the longitudinal (O) and transverse (□) velocity components. For comparison, a Gaussian distribution is illustrated as a dashed line. 116

5.18 Experimental mean frequency $\bar{f}(n)$ of each intrinsic function mode. An exponential law, e.g., $\bar{f}(n) \sim \beta^{-n}$ is observed with a scaling exponent $\beta = 2.01 \pm 0.02$, indicating a dyadic filter bank of the empirical mode decomposition. 117

5.19 Experimental joint probability density function $p(\omega, \mathcal{A})$ for a) the longitudinal velocity component, and b) the transverse one. A scaling trend, namely

	skeleton, e.g., $p_{max}(\omega) = p(\omega, \mathcal{A}_s(\omega)) = \max_{\mathcal{A}} \{p(\omega, \mathcal{A})|\omega\}$, $\mathcal{A}_s(\omega)$ is illustrated by a ○.	118
5.20	Experimental skeleton $\mathcal{A}_s(\omega)$. Power-law behavior is observed on the range $10 < \omega < 1000$ Hz with a Kolmogorov-like scaling exponent 0.33 ± 0.04. The inset shows the compensated curve to emphasize the observed scaling behavior.	118
5.21	Experimental skeleton $p_{max}(\omega)$. Power-law behavior is observed on the range $10 < \omega < 1000$ Hz with a Kolmogorov-like scaling exponent 0.67 ± 0.04. The inset shows the compensated curve to emphasize the observed scaling behavior.	119
5.22	Experimental marginal distribution of $p(\omega)$. Power-law behavior is observed on the range $10 < \omega < 3000$ Hz with a scaling exponent $\xi(0) \simeq -1$. The inset shows the compensated curve to emphasize the observed scaling behavior using the scaling exponent -1.	120
5.23	Experimental zeroth-order scaling exponent $\xi(0)$ obtained from different processes.	120
5.24	Experimental qth-order marginal moments $\mathcal{L}_q(\omega)$: a) and b) for the longitudinal velocity, c) and d) the transverse one. Power-law behavior is observed on the range $10 < \omega < 1000$ Hz.	121
5.25	a) Experimental scaling exponents $\xi(q) - 1$; b) the measured singularity spectrum $f(\alpha)$. For comparison, a lognormal formula with intermittency parameter $\mu = 0.30$ (solid line) and a fitted one $\mu = 0.18$ (dashed line) are also shown.	122
5.26	Illustration of a 0.05 second portion of temperature data $\theta(t)$ (thick solid line) as passive scalar turbulence. A typical ramp-cliff structure is observed with a gradually decreasing structure and a sharp increasing part. For comparison, a pure sine wave with 150 Hz is also shown. Visually, the ramp-cliff structure is significantly different with the sine wave.	124
5.27	a) Experimental Fourier spectrum $E_\theta(f)$. Power-law behavior is observed on the range $100 < f < 1000$ Hz with a scaling exponent 1.56 ± 0.03. The inset shows the compensated curve to emphasize the scaling behavior. b) Experimental second-order structure function $S_2^\theta(\tau)$, in which the Fourier-based inertial range $0.001 < \tau < 0.01$ seconds is illustrated by a solid line. The inset shows the compensated curve using the KOC value $2/3$.	125
5.28	a) Measured cumulative function $\mathcal{P}(\tau, f)$ and the corresponding co-cumulative function $\mathcal{H}(\tau, f)$ at two time scales $\tau = 0.002$ sec and $\tau = 0.01$ sec, corresponding to $f_o = 500$ Hz and 100 Hz in the inertial range. b) The corresponding $\mathcal{P}_1(\tau)$ and $\mathcal{H}_1(\tau)$. The inertial range $0.001 < \tau < 0.01$ sec predicted by the Fourier power spectrum is illustrated by a solid line.	126
5.29	a) Measured high-order structure functions $S_q^\theta(\tau)$. The inertial range $0.001 < \tau < 0.01$ sec predicted by the Fourier power spectrum is illustrated by a solid line. b) The extended-self-similarity plot of the qth-order $S_q^\theta(\tau)$ versus $S_3^\theta(\tau)$ on the inertial range.	127

5.30 a) Measured extended-self-similarity scaling exponents $\zeta_\theta(q)$ (○). For comparison, the scaling exponent provided by a lognormal formula with an intermittency parameter $\mu_\theta = 0.20$ (solid line), and the one compiled by Schmitt (2005) (□) are also shown. b) The corresponding singularity spectrum $f_\theta(\alpha_\theta)$. 127

5.31 a) Measured location τ_0 for the minimum value of the autocorrelation function $\rho_\tau(\ell)$ of temperature increments. The solid line indicates $\tau_0(\tau) = \tau$. b) Experimental $\Gamma_\theta(\tau)$, in which the Fourier inertial range is indicated by a solid line. Power-law behavior is observed on the range $0.0002 < \tau < 0.005$ sec, corresponding to a frequency range $200 < f < 5000$ Hz with a scaling exponent 0.86 ± 0.02. Inset shows the compensated curve using a fitted scaling exponent to emphasize the power-law behavior. 128

5.32 Measured maximum pdf $p_{\max}(\tau)$ (○) and zero-cross $p_z(\tau)$ (□). A double power-law behavior is observed for $p_{\max}(\tau)$ on the range $0.00004 < \tau < 0.0004$ sec, corresponding to a frequency range $2,500 < f < 25,000$ Hz, and $0.001 < \tau < 0.004$ sec, corresponding to $250 < f < 1,000$ Hz, respectively. The measured scaling exponents are 0.59 ± 0.01 and 0.83 ± 0.02. A single power-law behavior is observed for $p_z(\tau)$ on the range $0.00004 < \tau < 0.004$ sec, corresponding to a frequency range $250 < f < 25,000$ Hz, with a scaling exponent 0.45 ± 0.01. The inset shows the compensated curve to emphasize the observed scaling behavior. 129

5.33 a) Measured maximum first-order detrended structure function $\mathcal{B}_q(\tau)$. Power-law behavior is observed on the range $0.001 < \tau < 0.01$ seconds, corresponding to a frequency range $100 < f < 1000$ Hz. b) The extended-self-similarity plot $\mathcal{B}_q(\tau)$ versus $\mathcal{B}_3(\tau)$ on the range $0.001 < \tau < 0.01$ seconds. 130

5.34 a) Measured detrended structure function scaling exponent $\zeta_\theta(q)$ (○) and the extended-self-similarity scaling exponent $\zeta_\theta^R(q)$ (△). For comparison, a scaling exponent $\zeta_\theta(q)$ compiled by Schmitt (2005) is also shown as □. A lognormal with an intermittent parameter $\mu_\theta = 0.20$ is illustrated as a solid line. b) The corresponding singularity spectrum $f_\theta(\alpha_\theta)$. 131

5.35 a) Experimental qth-order wavelet leaders-based moments $\mathbf{Z}_q(\tau)$ on the range $0 \le q \le 6$. Power-law behavior is observed on the range $0.001 < \tau < 0.01$ sec. b) The corresponding singularity spectrum $f_\theta(\alpha_\theta)$. For comparison, the lognormal formula with an intermittency parameter $\mu_\theta = 0.20$ (solid line) and the compiled value (□) are also shown. The inset shows the measured scaling exponent $\xi_\theta(q)$. 132

5.36 a) Experimental probability density function of the extracted number of intrinsic mode functions. For comparison, a Gaussian distribution is illustrated by a solid line. b) The corresponding mean frequency $\overline{f}_\theta(n)$ of each intrinsic mode function. An exponential law is observed with a scaling exponent 1.81 ± 0.03, indicating a dyadic-like filter bank property of the Empirical Mode Decomposition. The inset shows the compensated curve to emphasize the observed exponential law. 132

5.37 Comparison of the second-order Hilbert marginal spectrum $\mathcal{L}_2^\theta(\omega)$ (○) and the Fourier power spectrum $E_\theta(f)$ (□). Power-law behavior is observed on the

List of figures xvii

range $100 < f < 1000\,\text{Hz}$ with scaling exponent $\xi_\theta(2) = 1.69 \pm 0.02$ and 1.56 ± 0.03, respectively for the Hilbert and Fourier approaches. The inset shows the compensated curve using the nonintermittent KOC value $2/3$. 133

5.38 a) Measured qth-order Hilbert marginal spectrum $\mathcal{L}_q^\theta(\omega)$ on the range $0 \le q \le 6$. Power-law behavior is observed on the range $100 < \omega < 1000\,\text{Hz}$. b) Experimental singularity spectrum $f_\theta(\alpha_\theta)$. For comparison, the lognormal formula with an intermittency parameter $\mu_\theta = 0.20$ (solid line) and the compiled value ($\square$) are also shown. The inset shows the measured scaling exponent $\xi_\theta(q)$. 134

5.39 The singularity spectra $f_\theta(\alpha_\theta)$ provided by different approaches. 135

5.40 a) Experimental Fourier power spectrum $E_L(f)$ of the Lagrangian $V_x(t)$. Power-law behavior is observed on the range $0.01 < f\tau_\eta < 0.1$ with a scaling exponent 1.88 ± 0.02. The inset shows the compensated curve using the nonintermittent Kolmogorov-Landau value 2. b) Measured second-order Lagrangian velocity structure function $S_2^L(\tau)$. The inset shows the compensated curve by using the nonintermittent Kolmogorov-Landau value 1. 137

5.41 a) Measured cumulative function $\mathcal{P}(\tau, f)$ (solid line) and the corresponding cocumulative function $\mathcal{H}(\tau, f)$ (dashed line) at two time scales $\tau/\tau_\eta = 10$ (thick solid line) and $\tau/\tau_\eta = 50$ (thin solid line), corresponding $f\tau_\eta = 0.1$ and 0.02 in the inertial range. b) The corresponding $\mathcal{P}_1(\tau)$ ($\bigcirc$) and $\mathcal{H}_1(\tau)$ ($\square$). The inertial range $10 < \tau/\tau_\eta < 100$ predicted by the Fourier power spectrum is illustrated by a solid line. 138

5.42 a) Experimental high-order Lagrangian structure functions $S_q^L(\tau)$, in which the inertial range predicted by the Fourier power spectrum is illustrated by a solid line. b) Extended-self-similarity plot $S_q^L(\tau)$ versus $S_2^L(\tau)$ on the range $3 \le \tau/\tau_\eta \le 100$. The vertical solid line indicates a separation scale $\tau/\tau_\eta = 30$. 139

5.43 a) Measured extended-self-similarity scaling exponent $\zeta_L(q)$. For comparison, the value predicted by the multifractal model of Biferale et al. (2004) (solid line) and the value from the experiment (denoted as Xu et al., $\triangle$) are also shown. b) The corresponding singularity spectrum $f(\alpha)$. 140

5.44 a) Experimental probability density function of the number of extracted intrinsic mode functions, in which the Gaussian distribution is illustrated by a solid line. b) Measured mean frequency $\bar{f}(n)$ for the first six modes. An exponential-law, e.g., $\bar{f}(n) \sim \gamma^{-n}$ is observed with a scaling exponent $\gamma = 1.75 \pm 0.05$, indicating a dyadic-like filter bank of the empirical mode decomposition. 141

5.45 Experimental Hilbert-based high-order moments $\mathcal{L}_q(\omega)$. Power-law behavior is observed on the range $0.01 < \omega\tau_\eta < 0.1$ for all q considered here. The inset shows the compensated curve using the fitted slope to emphasize the observed scaling behavior. 141

5.46 a) Experimental Hilbert-based scaling exponent $\zeta_L(q)$ ($\bigcirc$), in which the error-bar indicates 95% fitting confidence interval on the range $0.01 < \omega\tau_\eta < 0.1$.

xviii List of figures

b) The corresponding singularity spectrum $f(\alpha)$ versus α. For comparison, the experiment value provided by Xu et al. (2006a) is also shown as $\triangle$. 142

5.47 A vortex trapping event along a Lagrangian trajectory: a) Lagrangian velocity $V_x(t)$, b) the corresponding acceleration $a_x(t)$, and c) the energy dissipation rate $\epsilon(t)$ along this trajectory. The vortex trapping event is visible on the time span $80 < t/\tau_\eta < 120$ with a typical time scale $3 \sim 5\,\tau_\eta$. 143

5.48 A 100 sec of temperature measured in the near side wall of the Rayleigh-Bénard convection cell with a Rayleigh number $Ra = 1.31 \times 10^{10}$. A thermal plume, energetic structure is visible. 145

5.49 a) Experimental Fourier power spectrum $E_\theta(f)$ in a log-log plot. Power-law behavior is observed on the range $0.1 < f < 1$ Hz with a fitted scaling exponent $\simeq 1.20$. The inset shows a compensated curve using the fitted scaling exponent. b) Semi-log plot of $E_\theta(f)$ to emphasize the large-scale circulation with a typical frequency $f_L \simeq 0.03$ Hz, corresponding to a time scale $T_L = 33$ sec. The inset shows the measured autocorrelation function $\rho_\theta(\tau)$. 147

5.50 a) Experimental second-order structure function $S_2^\theta(\tau)$ for the case $Ra = 1.31 \times 10^{10}$. Power-law behavior is observed on the range $1 < \tau < 10$ sec with a fitted scaling exponent $\zeta_\theta(2) = 0.27 \pm 0.03$. The inset shows a compensated curve using the fitted scaling exponent. b) Measured cumulative function $\mathcal{P}_\theta(\tau,f)$ (solid line) and cocumulative function $\mathcal{H}_\theta(\tau,f)$ (dashed line) for separation time scale $\tau = 1$ sec (thick solid line) and $\tau = 5$ sec (thin solid line). 148

5.51 Experimental cumulative function $\mathcal{P}_1(\tau)$ ($\bigcirc$) and co-cumulative function $\mathcal{H}_1(\tau)$ ($\square$). The expected power-law range $1 < \tau < 10$ sec is illustrated by vertical lines. 148

5.52 Experimental high-order structure functions $S_q^\theta(\tau)$: a) $Ra = 1.31 \times 10^{10}$, and b) $Ra = 9.5 \times 10^9$. Power-law behavior is observed on the range $1 < \tau < 10$ sec. 149

5.53 a) Experimental scaling exponent $\zeta_\theta(q)$. b) The corresponding singularity spectrum $f_\theta(\alpha_\theta)$ versus α_θ. 149

5.54 Experimental $\Gamma_\theta(\tau)$. Power-law behavior is observed on the range $1 < \tau < 10$ sec with a scaling exponent 0.42 ± 0.02. The inset shows the compensated curve using the fitted scaling exponent. 150

5.55 Experimental $p_{\max}^\theta(\tau)$. Power-law behavior is observed on the range $1 < \tau < 10$ sec with a scaling exponent 0.33 ± 0.02. The inset shows the compensated curve using the fitted scaling exponent. 151

5.56 A snapshot of the vorticity field $\omega(x,y)$ obtained using a very high resolution direct numerical simulation on the range $0 \le x,y \le \pi/2$ (left) and $0 \le x,y \le \pi/4$ (right). High intensity vorticity events are discretely distributed in space with a typical wavenumber $k \simeq k_f = 100$, corresponding to 80 grid points, approximately. 155

5.57 a) Measured Fourier power spectrum $E_\omega(k)$, in which the forcing scale $k_f = 100$ is illustrated by a vertical solid line. The inset shows a compensated

	curve using a scaling exponent -2. b) The corresponding second-order structure function $S_2(\ell)$, in which the power-law range is indicated by a vertical solid line. The inset shows the local slope $\zeta_\omega(2,\ell)$.	155
5.58	a) Measured cumulative function $\mathcal{P}_\omega(\ell,k)$ and the corresponding co-cumulative function $\mathcal{H}_\omega(\ell,k)$ with two separation scale $\ell = 0.001$ (thin solid line) and $\ell = 0.002$ (thick sold line) lying in the forward enstrophy cascade. b) Measured $\mathcal{P}_1(\ell)$ ($\bigcirc$) and $\mathcal{H}_1(\ell)$ ($\square$). The forward enstrophy range is indicated by a vertical line.	156
5.59	a) Extended-self-similarity plot of the high-order structure functions $S_q(\ell)$ versus $S_2(\ell)$ in the forward enstrophy cascade range $0.001 < \ell < 0.01$. b) Measured singularity spectrum $f(\alpha)$ versus α. The inset shows the measured scaling exponent $\zeta_\omega(q)$, in which the dashed line indicates $q/2$.	157
5.60	a) Experimental probability density function $p(N)$ of the number of extracted intrinsic mode functions, in which the Gaussian distribution is illustrated by a solid line. b) Measured mean wave number $\bar{k}(n)$ for the first 11 modes. Exponential-law is observed with a scaling exponent $\gamma = 1.98 \pm 0.02$, indicating a dyadic filter bank of the empirical mode decomposition.	158
5.61	a) Measured second-order Hilbert moment $\mathcal{L}_2(k)$. A dual-power-law behavior is observed on the range $3 < k < 20$ for the inverse energy cascade, and $200 < k < 2000$ for the forward enstrophy cascade. b) The corresponding compensated curve using the fitted scaling exponent to emphasize the scaling behavior.	158
5.62	a) Measured high-order Hilbert moment $\mathcal{L}_q(k)$ for q between 1 and 4 from bottom to top. b) The corresponding measured scaling exponent $\zeta_\omega^F(q)$ ($\bigcirc$) for the forward enstrophy cascade and $-\zeta_\omega^I(q)$ ($\square$) for the inverse energy cascade. For comparison, the solid line is $q/3$ and the dashed line is for a log-Poisson fitting $\zeta_\omega^F(q) = q/3 + 0.45(1 - 0.43^q)$.	159
5.63	Hilbert-based singularity spectrum $f(\alpha)$ for the forward enstrophy cascade ($\bigcirc$), and the inverse energy cascade ($\square$). For comparison, the log-Poisson based one is illustrated as a solid line, and the structure function based one for the forward cascade is illustrated by $\triangle$.	160
6.1	A 24 hours portion of the U-velocity time series, superposed on the height information, indicating the tidal range.	164
6.2	Power spectrum of the three components of the ADV velocity, superposed to the water level power spectrum. A power-law fit $f^{-0.58}$ is superposed on U and V power spectra of the velocity.	164
6.3	Arbitrary order Hilbert spectral analysis of the U-velocity data in the bottom boundary layer: moments of order 0 to 4 are displayed. The scaling exponents for scales from $12\,s$ to 1 hour are estimated.	165
6.4	The generalized moment function $\xi(q)$ estimated for the U velocity data in the bottom boundary layer, for scales from $12\,s$ to 1 hour. The inset shows the singularity spectrum.	166
6.5	A 14 days portion of the modelled water-level time series at Dunkirk, and the measured data (above, superposition of the model and measurements). The	

	time series difference (measurements-model) shows that the superposition is not perfect (bottom).	167
6.6	A 6 months portion of the difference time series at Dunkirk (measurements-model), showing its stochastic and multiscale dynamics.	167
6.7	The pdf of the difference data, computed over the whole time series, with two exponential fits, of equation $p(x) = A \exp(bx)$, with $A = 0.52$ and $b = 7.7$ for $x < 0$ and $A = 0.37$ and $b = -6.2$ for $x > 0$.	168
6.8	Left: Power spectra of model and measurements of water-level data (from 23 April 2003 to 7 August 2007). The superposition is good only for high-frequency cycles. For scales larger than one day the superposition is not good, and the model does not reproduce the dynamics revealed by the data. Right: This is highlighted by the ratio of spectra, which is mostly between 100 and 1,000, indicating that between one day and six months the natural variability is much larger than modeled.	169
6.9	Scaling of the moments using Hilbert spectral analysis of the water-level time series.	170
6.10	a) The moment function $\xi(q)$ estimated for the stochastic part of water-level fluctuations, for scales between 1 and 25 days. b) the associated singularity spectrum.	170
6.11	Presentation of the 4 MAREL time series considered here. a) Time series of air and water temperature. There is an annual cycle in both cases. The temperature difference (air-water) is also shown. b) The dissolved oxygen (percentage of saturation) and fluorescence time series. The oxygen does not display a clear annual cycle; the fluorescence has an annual spring bloom.	172
6.12	a) Hilbert spectral analysis, for the moment $q = 1$ of air, water temperature and the difference. Power fits are also shown, corresponding to $H = 0.2$ and 0.4 for the air and water temperature series, respectively. b) The probability density function of the difference $\delta T = T_A - T_O$ The dotted line is the symmetric. It shows that the pdf is not symmetric and is roughly an exponential.	173
6.13	Hilbert spectral analysis, for the moment $q = 1$ of dissolved oxygen and fluorescence data. Power fits are also shown, corresponding to $H = 0.42$ and 0.5 for the dissolved oxygen and fluorescence series, respectively.	174
6.14	Scaling of Hilbert spectral analysis for the dissolved oxygen (left) and fluorescence data (right). The scaling range considered is between the daily and annual cycles.	175
6.15	Left: Scaling exponents $\zeta(q) = \xi(q) - 1$ of both dissolved oxygen and fluorescence data. There is a much stronger intermittency for the latter series. Right: Associated singularity spectrum.	175
6.16	A one-hour portion of the atmospheric wind velocity measurements.	176
6.17	a) Power spectrum of the wind velocity measurements, where a power-law range is found on the range 10^{-4} to 10 Hz with a slope of 1.27 ± 0.04.	

List of figures

6.18 a) The structure functions applied on the wind velocity series, for moments from 1 to 6; b) the compensated structure functions, showing the scaling range, from 5 to 500 seconds. ... 177

b) The compensated spectrum, showing that the power-law is found precisely for frequencies on the range 10^{-4} to 10^{-1} Hz. ... 177

6.19 a) The scaling exponent $\zeta(q)$ estimated from the wind velocity time series using structure functions, for scales from 5 to 500 seconds; b) the associated singularity spectrum. ... 178

6.20 a) Arbitrary order Hilbert spectral analysis applied on the wind velocity data, for moments from 0 to 5; the scaling range is confirmed by the compensated values in b). ... 178

6.21 a) The scaling exponents $\zeta(q) = \xi(q) - 1$, obtained from the arbitrary order Hilbert spectral analysis of wind velocity, with a fit over the range 0.003 to 1 Hz; b) the associated singularity spectrum. ... 179

6.22 A typical transfer function from the wind velocity to the wind power, with arbitrary numerical values given here as illustration. It should be seen as a conditional average – instantaneous values do not follow this curve. There are four zones: below a given threshold (here 3 m/s), there is no production since the velocity is too low. For the increasing part, the power is proportional to the cube of the velocity; then around a given value (here 10 m/s) there is a plateau and for excessive values of the velocity (here 16 m/s), for safety reasons the windmill is not operating and the production goes to zero again. ... 180

6.23 The power output time series studied here, showing its high multiscale variability, and also some periods when the wind farm is stopped, when the wind input is below a given threshold. Only the first part of the time series is studied here. ... 181

6.24 a) Power-law scaling of the output power series, with a slope of 1.75 ± 0.04, close to 5/3; b) compensated spectrum showing the power-law range. ... 181

6.25 a) The structure functions applied on the wind power series, for moments from 1 to 6; b) the compensated structure functions, showing the scaling range, which is quite large, over almost 4 orders of magnitude. ... 182

6.26 a) The scaling exponent $\zeta(q)$ estimated from the wind power time series using structure functions, for scales from 1 to 5,000 seconds. b) The associated singularity spectrum. ... 182

6.27 a) Arbitrary order Hilbert spectral analysis applied on the wind power data, for moments from 0 to 5. There is a nice scaling, confirmed by the compensated values in b). ... 183

6.28 a) The scaling exponents $\zeta(q) = \xi(q) - 1$, obtained from the arbitrary order Hilbert spectral analysis, with a fit on the range 10^{-4} to 0.2 Hz. b) The associated singularity spectrum. ... 183

Preface

Fluid geoscience, which includes the study of the atmosphere, the oceans, hydrology, and climate, is a special topic in science: it is the application, in some sense, of the Navier-Stokes and transport equations, as well as other fluid mechanics theoretical concepts, into the "dirty real-world". In the theoretical realm and in the laboratory, where all conditions are controlled, a desired phenomenon can be isolated and studied. In the field, however, conditions are not controlled, leading to a multitude of mixed phenomena, and measurements are the result of a multiscale complex system. In such a system, stochastic fluctuations are often superposed onto deterministic forcing associated with astronomic cycles. Scaling concepts are very important when approaching and studying geophysical fields, but the stochastic scaling properties are perturbed by forcing of, say, the day-night cycle, the annual cycle, or the tidal cycle in oceanography. In such cases, scaling methodologies may fail. In this book, we shall discuss such issues, and provide new tools that are more adapted to retrieving stochastic scaling information from different turbulence and real-world situations.

Fluid geoscience is also a special topic, in the sense that it is extremely difficult to find adequate and solid theories. The underlying fundamental equations (e.g., Navier-Stokes) are nonlinear, and turbulence belongs to the last field of classical physics, which is still unsolved. The resolution of Navier-Stokes equations belongs to the Clay Institute's one-million-dollar problems. This explains why the field of fluid geosciences has only progressed, in some sense, so slowly. Nowadays, many models used are based on crude large-scale approximations of dynamics, based on averages of Navier-Stokes equations. These models are inaccurate and no theory really tackles the nonlinearities and intermittencies of the natural (fluid) turbulent-like world. New theories and new models are needed. The solution may reside in generalizations and continuations of Kolmogorov's seminal 1941 work, when a scaling law was proposed to describe velocity fluctuations at different scales. Multifractal models are continuations of these works from the 1940s, and their aim has been to find statistical laws (such as thermodynamics in statistical physics for an equilibrium system) able to describe and model, at different scales, the fluctuations of fields of interest.

Such a general objective explains why we have devoted the first chapters of this book to historical developments on scaling and multiscaling issues in the field of turbulence and passive scalar turbulent transport. We have then followed this with an overview of several scaling methods, including our own results and proposals. After these methodological presentations, we have given the remaining chapters to emphasising applications in the fields of turbulence, and the real worlds of the ocean and the atmosphere.

Many more applications are possible, and it is with this aim, and to help other colleagues willing to develop their own approach and compare their results with other methodologies, that we have installed for free download on a dedicated website most of the Matlab codes used in this book.

We thank Rudy Calif, Enrico Calzavarini, Pierre Chainais, Laurent Chevillard, Gaël Dur, Patrick Flandrin, Yves Gagne, Guowei He, Norden E. Huang, Shaun Lovejoy, Nicolas Perpete, Daniel Schertzer, Sami Souissi, Federico Toschi, Lipo Wang, Ke-Qing Xia, and Quan Zhou for discussion on these topics in the recent last years.

The providers of the data used in the book are acknowledged: We thank Professor Charles Meneveau for sharing his experimental velocity database, which is available for download at his web page: www.me.jhu.edu/meneveau; Professor Yves Gagne for providing us the temperature data as the passive scalar turbulence; Professor Ke-Qing Xia for providing us the temperature data obtained from the Rayleigh-Bénard convection cell; and Professor Guido Boffetta for sharing his high-resolution numerical data with us, which is available at http://cfd.cineca.it. SHOM is acknowledged for the water-level data set. Renosh P. R. provided the ADV in situ coastal velocity data set and helped in doing some analyses. Alain Lefebvre and IFREMER are acknowledged for the MAREL data set. Rudy Calif made available to us the atmospheric and wind energy data set. The EMD Matlab codes used in this paper were written by Dr. Gabriel Rilling and Professor Patrick Flandrin from Laboratoire de Physique, CNRS & ENS Lyon (France): http://perso.ens-lyon.fr/patrick.flandrin/emd.html, and modified by Yongxiang Huang.

We have met many colleagues, and exchanged many ideas, in several conferences in the past years, such as European Turbulence conferences, American Geophysical Union general assemblies in San Francisco and European Geoscience Union general assemblies in Vienna, in the Nonlinear Geophysics section and the Nonlinear Processes in Geosciences division, respectively.

In writing this book, we have received financial support from several institutions, such as the National Science Foundation of China (NSFC with grant numbers 11202122 and 11332006); the Fundamental Research Funds for the Central Universities (Grant No. 20720150075); the French National Centre for Scientific Research (CNRS), University of Lille 1, University of Littoral (ULCO); the Laboratory of Oceanology and Geosciences (LOG); the Shanghai Institute of Applied Mathematics and Mechanics, and Shanghai University, as invited professors in China or in France. We thank these institutions for their continuing support.

Peter Magee (www.englisheditor.webs.com) is also thanked for his English proofing of the manuscript.

Finally, we would like to dedicate this book to our families.

1

Introduction: a multiscale and turbulent-like world

1.1 Data from the real world

Data analysis is an essential part of scientific research and engineering applications. A proper data analysis method will provide a better understanding of the process under consideration, for example, extracting useful parameters to validate an existing theory or to inspire a new theory. However, data from the real world, such as field-observation data, well-controlled laboratory experiments, or numerical simulation, generally possess several problems. For instance, the data length may be too short to satisfy stationary or ergodic conditions, or the mechanism behind the data may be nonlinear, etc. In Figures 1.1 through 1.3 are several examples from the real world, demonstrating common problems of real datasets.

Figure 1.1a displays a collected oxygen saturation index obtained from a MAREL network (Automatic Monitoring Network for Littoral Environment, Ifremer, France) from the period January 1 to December 31, 2010. The large variation of the measured oxygen saturation index shows the nonstationarity of the data. For example, a high intensity of oxygen saturation index was observed at September 3, 2010. As shown in Section 1.3, to mimic this nonstationary event, high-order Fourier harmonic components are required. Moreover, sometimes the sensor fails to collect data, due to maintenance problems or failure of the system. This missing data problem is typical of field observation data. To emphasize this, we replotted a 10-day portion of the data in Figure 1.1b, which shows a discontinuous curve due to the problem of the data missing. This imposes a difficulty for the Fourier-based data analysis method, for which a uniform time step is often required. To see the irregular time step more clearly, the time interval δt has been shown between two successful measurements in Figure 1.1c. Visually, the time step δt demonstrates a strong intermittent distribution with large values of δt. Another aspect of this data set is that the underlying physical mechanism is nonlinear. It means that if one can write a governing equation for the oxygen

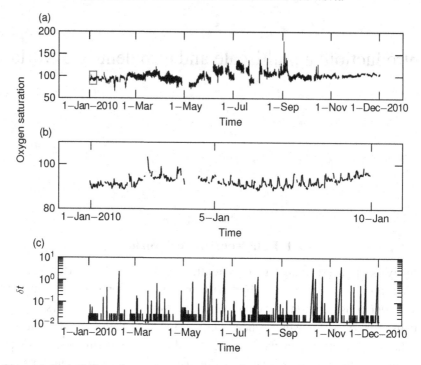

Figure 1.1 a) Collected oxygen saturation index on the time period January 1 to December 31, 2010, with a sampling time of 20 minutes. The data were collected by a sensor belonging to the MAREL network (Automatic monitoring network for littoral environment, Ifremer, France). Due to several reasons, e.g., failure of the sensor, there are several missing data points. A strong event was also observed around September 1, 2010, showing the nonstationarity of the process. b) An enlargement part on the time period January 1 to January 10, 2010. c) The time interval δt between two successful measurements. The intermittent distribution of δt demonstrates the problem of missing data or irregular time step.

saturation index, this equation must be nonlinear. Therefore, the difficulties of this type of data set are missing data/irregular time steps, both nonstationarity and nonlinearity.

Turning to another aspect of the real world data, namely varying sample size, considered here is an example from the global drifter program. Figure 1.2a illustrates several Lagrangian trajectories obtained from the global drifter program. Due to failures of the devices, battery life-time, etc., the life-time T of drifters varies from one to another. They provide measured physical quantities with different lengths. For example, the drifter collects data every six hours, corresponding to four data points per day. Therefore, a finite life-time T means a finite data sample, for example, $L = 4T$. Figure 1.2b shows the measured distribution of T. It has an exponential distribution, indicating a large variation of the data length.

1.1 Data from the real world

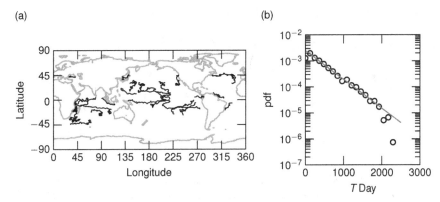

Figure 1.2 a) Illustration of the Lagrangian drifter trajectories obtained from the global drifter program. b) Measured pdf of the drifter life-time T. The measured pdf $p(T)$ has an exponential distribution. The drifter collected several physical quantities of the ocean with a sampling time of six hours. This provided a large variation of the data length.

More precisely, the mean and standard deviation of T are $T \simeq 370$ and $T \simeq 360$ days. This will impose difficulties when performing traditional spectrum analysis, since each drifter covers different time scales. This happens also in well-controlled laboratory experiments. For example, in Lagrangian turbulence experiments, the tracer particles are tracked experimentally in a turbulent flow with thousands of realizations (Toschi and Bodenschatz, 2009). Due to the finite measurement volume and the limitation of the particle tracking technique, the tracking period for each individual particle is different. We therefore have the same difficulty as the one for Lagrangian drifters.

The presence of a forcing scale, or strong deterministic forcing, is another important feature of the data sets from natural sciences, especially geophysics. For example, in geoscience, daily cycle and annual cycle are present in several collected data, and in marine data sets, the tidal cycle is also present for many series. This is typically the case for sea surface temperature, air temperature, river daily discharge, etc. Figure 1.3 displays a collected water temperature $\theta(t)\,°C$ from the MAREL network mentioned earlier in this chapter. The collected temperature possesses a strong annual cycle, a deterministic forcing from the solar-earth system. As shown in Chapter 4, the scaling behavior (see discussion in next subsection and more detail in Chapter 3), an important feature of complex systems, will be perturbed by such deterministic forcing. There exists a continuous range of excited time scales, at least between the annual cycle and the daily cycle, namely multiscale fluctuations. One of the objectives of this analysis is to try to better understand the statistical relationship between scales. Whether or not this relationship can be experimentally extracted from the data is one of the essential contents of this book. Some general

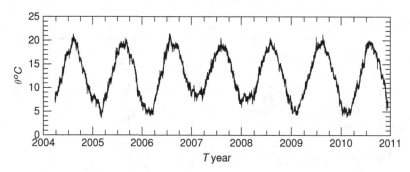

Figure 1.3 Collected water temperature data on the period March 24 to December 13, 2010. These data are automatically recorded every 20 minutes by the MAREL network. A strong annual cycle is observed. Note that this deterministic-like forcing is a typical structure in the geophysical data, which might induce difficulties in the scaling analysis.

comments on this issue are proposed in the next section, and more details are provided on this topic in the rest of this book.

Summarized in this list are the main properties of the data collected from the real world. They are:

1. finite sample size;
2. nonstationarity and nonlinearity;
3. presence of stochastic fluctuations, as well as deterministic forcing;
4. multiscale, or multiscaling statistical properties;
5. irregular time step, or missing data.

An ideal time-series analysis method should be able to handle all difficulties imposed by these the above listed properties to reveal the physics represented by the data.

1.2 Multiscale phenomena

As mentioned in Section 1.1 multiscale fluctuations are relevant in many complex systems in which many time or spatial degrees of freedom are present and may interact with each other. The most classical complex system is the case of turbulent flows (Frisch, 1995; Tsinober, 2009). In three-dimensional turbulent flows, a large range of time and spatial scales are involved and interact with each other, at least in the so-called inertial range. The famous Richardson-Kolmogorov energy cascade picture has been proposed to interpret the turbulent flow in a phenomenological way: the energy is transferred from large- to small-scale structures, until the viscosity scale, where the energy is converted into heat (see Kolmogorov theory in Chapter 2 or Frisch, 1995). Figure 1.4a displays a one second portion of the longitudinal

1.2 Multiscale phenomena

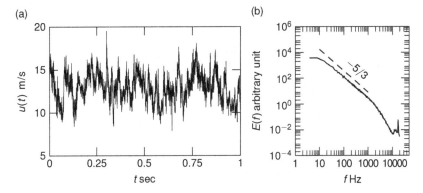

Figure 1.4 a) A one second portion of the measured longitudinal turbulent velocity in a wind tunnel experiment at Johns Hopkins University with a Taylor microscale based Reynolds number $Re_\lambda \simeq 720$. The measured velocity is fluctuating over a large range of time scales. b) The experimental Fourier power spectrum in a log-log plot. Power-law behavior $E(f) \sim f^{-\beta}$ is observed with a β close to the Kolmogorov 1941 theory prediction 5/3.

velocity obtained from a wind tunnel experimental with a Taylor's microscale based Reynolds number $Re_\lambda \simeq 720$. This experimental Eulerian velocity fluctuates over different time scales. High intensity events, known as nonstationary events, were observed. The corresponding Fourier power spectrum $E(f)$ is shown in Figure 1.4b in a log-log plot, in which a 5/3 power-law relation from the Kolmogorov 1941 theory is illustrated by a dashed line. A power-law behavior is observed on the range $10 < f < 1000$ Hz. It corresponds with the Kolmogorov's 1941 phenomenological theory (see more discussion in Chapter 2).

Another example is the Lagrangian velocity obtained from a high-resolution numerical simulation. Figure 1.5a shows the Lagrangian velocity along a Lagrangian trajectory and 1.5b the corresponding energy dissipation rate $\epsilon(t)/\langle\epsilon\rangle$. This trajectory was chosen on purpose to display a "vortex trapping" event around $t/\tau_\eta = 100$. The corresponding energy dissipation is fluctuating over a large range of amplitude. For the present trajectory, it is as high as 23 times the mean energy dissipation rate, showing a strong intermittent event. In fact, the highest energy dissipation rate could reach more than 100 times of the mean energy dissipation rate. Note that a continuous range of frequencies/scales is observed. The Richardson-Kolmogorov cascade picture is essentially a multiscale description of turbulent flows (Tsinober, 2009).

Hydrodynamic turbulent flows are governed by the Navier-Stokes equation (see Chapter 2). Unfortunately, a mathematical solution is unreachable for the moment. Therefore, a phenomenological theory, such as the Kolmogorov 1941 theory, must be verified experimentally or numerically. In the content of multiscale analysis, the

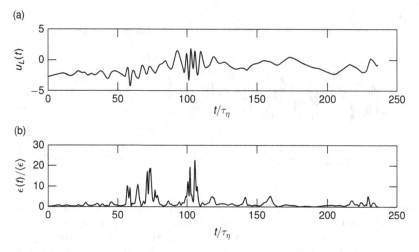

Figure 1.5 a) An example of turbulent velocity along a Lagrangian trajectory obtained from a high-resolution direct numerical simulation with a Taylor microscale-based Reynolds number $Re_\lambda \simeq 400$. A vortex trapping event is observed around $t/\tau_\eta = 100$. b) The measured energy dissipation rate $\epsilon(t)/\langle\epsilon\rangle$ along the same Lagrangian trajectory, showing a strong intermittent/nonstationary event around time $50 < t/\tau_\eta < 120$.

essential job of the data analysis is thus to reveal the statistical relations between different scales to verify theory predictions or to provide "seeds" for new theories.

There also exists numerous turbulent-like complex systems, such as financial activities (Ghashghaie et al., 1996; Mantegna and Stanley, 1996; Schmitt, Schertzer, and Lovejoy, 1999; Li and Huang, 2014); environmental variables in the sea (Schmitt et al., 2009; Zongo and Schmitt, 2011; Huang, Schmitt, and Gagne, 2014); daily river discharge (Tessier et al., 1996; Dahlstedt and Jensen, 2005; Huang et al., 2009a), etc., to name but a few. Unfortunately, an exactly governing equation for such a complex system cannot be written down. For now, various time series/data from field observations or well-controlled laboratory experiments have been obtained. A proper multiscale treatment of the data will provide a better understanding of the dynamics of such complex systems in order to extract the scale-dependent information/parameters. New theoretical considerations might be inspired by the data.

1.3 The Fourier-based methodology and its potential shortcomings

1.3.1 Linear asymptotic approximation

There are many time-frequency analysis methods (Cohen, 1995; Flandrin, 1998). Their basic idea originated in part from the Fourier analysis. It can be interpreted

1.3 The Fourier-based methodology and its potential shortcomings

as representing a given signal/function $x(t)$, by a given basis φ, i.e.,

$$x(t) = \int_{v \in R} \int_{t' \in R} \psi(t', v) \varphi(t, t', v) dv dt' \tag{1.1}$$

where ψ is a coefficient (function) that can be determined as:

$$\psi(t, v) = \int_{t' \in R} x(t) \varphi(t, t', v) dt' \tag{1.2}$$

Here, the basis function φ also can be interpreted as an integral kernel of Equation 1.2 (Cohen, 1995). It is an asymptotic approximation: the signal is asymptotically approximated by the chosen basis (function) φ. Usually, the property of the chosen basis is well known. Then the given signal is checked to see what it looks like with respect to the chosen basis (function) φ. For example, when the trigonometric function is chosen, the classical Fourier transform is obtained, for which ψ depends only on the frequency:

$$\psi(f) = \int_{-\infty}^{+\infty} f(x) e^{i 2\pi f x} dx \tag{1.3}$$

in which $\psi(f)$ is the Fourier coefficient. $\psi(f)$ is independent with x, corresponding to a global property of x. Therefore, the Fourier analysis cannot identify a nonstationary event (Cohen, 1995; Flandrin, 1998; Huang et al., 1998; Huang, 2009).

Another example is the Wavelet transform, where ψ depends on time t and the scale a:

$$\psi(a, t) = \frac{1}{\sqrt{a}} \int_{t' \in R^n} x(t') \varphi\left(\frac{t' - t}{a}\right) dt' \tag{1.4}$$

where n is the dimension of the space, $\varphi(t)$ is the so-called mother wavelet and a is a dilatation parameter. To be a mother wavelet, $\varphi(t)$ should satisfy some conditions. (For details on wavelet theory, see Daubechies (1992); Meyer (1995); Mallat (1999)). The wavelet transform approach may also be considered as an adaptive-window Fourier transform (Huang et al., 1998). Note that the wavelet transform is local on different scales. Therefore, it is efficient for detecting and analyzing nonstationary events. However, as shown in Chapter 4, the wavelet-based method is still affected by the high-order harmonic problem. Indeed, this problem cannot be overcome if the basis is chosen *a priori*.

In general, the traditional approach for time-frequency analysis is to choose basis functions *a priori*. Once the basis (function) is fixed, the information that can be extracted from the data is determined. They are also energy-based approaches: only when the event contains enough energy can it then be detected by such methods (Huang et al., 1998; Huang and Wu, 2005).

1.3.2 High-order Fourier harmonic: mathematical or physical?

To show experimentally the problem of high-order harmonic components, here the Fourier analysis of the nonlinear Duffing equation is considered. It is shown that the high-order harmonic component is a requirement of the mathematical approach used here. They are not observed in the physical domain.

The nonlinear Duffing equation is written as

$$\frac{d^2x}{dt^2} + x(1+\epsilon x^2) = b\cos(2\pi\Omega t) \qquad (1.5)$$

in which ϵ is a nonlinear parameter. It can be considered as a pendulum with forcing function $b\cos(\Omega t)$, in which its pendulum length varies with the angle, showing a nonlinear mechanism in this system through the parameter ϵ. The parameters chosen in Equation (1.5) are $\epsilon = 1$, $b = 0.1$ and $\Omega = 0.04$. Note that $\epsilon = 1$ is finite, and thus Equation (1.5) presents a strong nonlinearity without analytical solution. A fifth-order Runge-Kutta algorithm is performed to solve Equation (1.5) with an initial condition $[x(0), x'(0)] = [1, 1]$ and a sampling frequency of 10 Hz. Figure 1.6a shows the numerical solution. Some comments on the numerical solution are provided in this section. Firstly, visually the wave profile of the numerical solution $x(t)$ is far from a sine or cosine wave; this is a nonlinear distortion (Huang et al., 1998, 2011a). This deviation is a result of the nonlinear mechanism. Secondly, the solution $x(t)$ is smooth with a mean period $T \simeq 9.524$, provided by the peak counting method. It corresponds to a dominant frequency $f_D \simeq 0.105$ Hz. Therefore, a

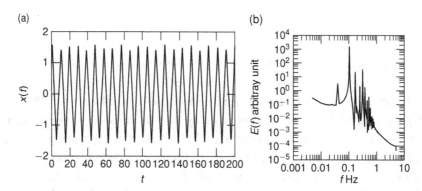

Figure 1.6 a) Numerical solution of the Duffing equation with $\epsilon = 1$, $b = 0.1$, $\Omega = 2\pi/25$ and initial value $[x(0), x'(0)] = 1$. Note that the solution $x(t)$ is smooth, with a mean period $T = 9.524$, corresponding to a domain frequency 0.105 Hz. Hence a frequency $f \gg 0.105$ is unphysical. b) The corresponding Fourier power spectrum. The forcing scale is observed at 0.04 Hz as the first peak in $E(f)$. The second peak is at 0.105 Hz for the domain frequency. High-order Fourier harmonic components are visible for $f \gg 0.105$ Hz. They are required by the linear approximation of the Fourier transform.

1.3 The Fourier-based methodology and its potential shortcomings

frequency much higher than this value is artificially found by the analysis method: this is a "high-order harmonic problem" (Cohen, 1995; Flandrin, 1998).

Figure 1.6b shows the measured Fourier power spectrum in a log-log plot. A strong peak is observed at $f \simeq 0.1$, which agrees well with the dominant frequency f_D in the previous paragraph. Also noted were the existence of several peaks. The first peak at $f \simeq 0.04\,\text{Hz}$, corresponds to the external forcing Ω. The other peaks are larger than f_D and thus considered as high-order harmonic components. They are required by the Fourier analysis. This phenomenon is also observed for the Wavelet transform; for more details, see Huang et al. (2011a).

To show the high-order harmonic more clearly, $x(t)$ was reconstructed partially from several Fourier coefficients, i.e., $\tilde{x}(t) = \int_{f \in F} \mathcal{F}(f) \exp(-j2\pi f t) df$ for a given range of $f \in F$. Figure 1.7 shows (a) the reconstruction $\tilde{x}(t)$ from the domain

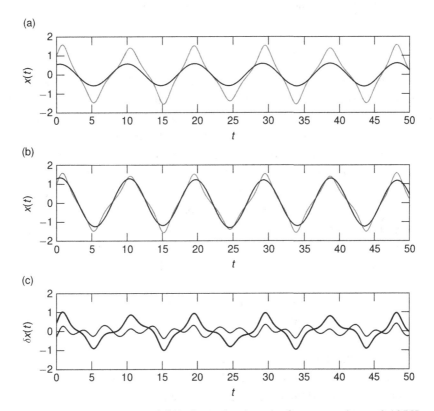

Figure 1.7 a) Reconstructed $\tilde{x}(t)$ from the domain frequency $f_D = 0.105\,\text{Hz}$. b) Reconstructed $\tilde{x}(t)$ from the first 30 Fourier modes. c) Residue $\delta x(t) = x(t) - \tilde{x}(t)$, in which the thick solid line is for the domain frequency and the thin line is for the first 30 Fourier modes. Note that the high-order Fourier harmonic components are the result of the difference between the Duffing solution $x(t)$ and the sine wave profile.

frequency f_D, and (b) the reconstruction from the first 30 Fourier modes (resp. up to $f \simeq 0.15\,\text{Hz}$, the first valley in between $0.1 < f < 0.2\,\text{Hz}$). As mentioned here, the idea of the Fourier analysis is to approximate a given signal by using a series of sine or cosine waves. Therefore, with an increasing number of Fourier modes, the reconstructed signal becomes closer and closer to the original one. Figure 1.7c displays the residue $\delta x(t) = x(t) - \tilde{x}(t)$, in which the thick solid line is for the dominant Fourier mode and the thin solid line is for the first 30 Fourier modes. This clearly shows that due to the so-called nonlinear distortion, the high frequency oscillation (resp. the so-called high-order harmonics) is emerging in $\delta x(t)$. However, this high-frequency oscillation is not observed in the original signal $x(t)$. The high-order harmonic component is thus a requirement of the Fourier analysis, but is not a physical requirement. It will be shown in Chapter 4 that by introducing the idea of intrawave frequency modulation, the high-order harmonic component is then constrained by the use of a Hilbert-based method.

1.4 Conclusion and further remarks

In conclusion of this introductory chapter, for several reasons, the data obtained from the real world, natural sciences, and geosciences in general, are typically nonstationary and nonlinear. The data also display multiscale properties with fluctuations over a large range of scales. An ideal data processing method should handle all aspects mentioned in the previous section: nonstationary, nonlinear, multiscale, irregular time step/missing data, etc. In recent years, several methodologies have been proposed to handle such real data. These include the Wavelet transform, Detrended Fluctuation Analysis, Empirical Mode Decomposition, and the associated Hilbert-Huang Transform. They have been successful in some aspects, but in others, they have had difficulties. For example, the Wavelet transform is efficient for nonstationary processes, but it might be affected by nonlinear processes (Huang et al., 1998, 1999).

Therefore, before a method is applied to data from the real world, it is better to know whether this method is suitable for the given type of data. For example, a strong annual cycle is often observed in oceanic time series (see Figure 1.3). Due to the presence of the annual cycle, the structure function analysis is strongly influenced when the multiscaling/multifractal property of such data set is concerned. The result is then biased (Huang et al., 2011a); see also discussion in Chapter 4.

In this book, the theory of homogeneous and isotropic turbulence in Chapter 2 is first recalled as the most classical multiscale complex system. Here the ideas of energy cascade, intermittency, and multifractal are briefly described, along with discussions on several intermittency models. Chapter 3 presents several scaling stochastic processes, including self-similar processes and nonstationary

1.4 Conclusion and further remarks

and more recent multifractal random walks. Several relevant methodologies to detect the scaling properties are presented in Chapter 4. They include the classical structure function analysis, detrended fluctuation analysis, detrended structure function, autocorrelation function scaling of increments, maximum probability density function scaling of increments, wavelet-based approaches, empirical mode decomposition, and the associated Hilbert spectral analysis, where their advantages and potential shortcomings are discussed in detail. In Chapters 5 to 6, applications from turbulence and geophysics are presented.

2
Homogeneous turbulence and intermittency

2.1 Introduction

In this book, we will consider stochastic time series with scaling properties and how to analyze them. The general inspiration, and the theoretical foundation of our approach, is the field of fully developed turbulence. Many properties of fields and time series, and many concepts, have been found and developed in the field of turbulence, and later exported to other complex fields such as engineering and natural sciences. Hence in this introduction chapter, we have revised some basic concepts and results in the field of locally homogeneous and isotropic three-dimensional turbulence. Older, more general references in this respect are the following: Batchelor (1953); Monin and Yaglom (1971); Tennekes and Lumley (1972); Frisch (1995); Pope (2000). With our consideration of time series, turbulence is first presented as a 3D problem, from which time series are extracted, and which inherit mainly nonlinear properties from the 3D world.

Turbulence is a phenomenon that is still only partly understood. However it concerns directly or indirectly a large number of scientific fields: meteorology, oceanography, astronomy, hydraulics, aeronautics, environment, chemical industry, automotive engineering, renewable energy, etc. A better understanding and theoretical description of turbulence is badly needed for practical applications in all these fields.

The first use of "turbulence" for sinuous motions in fluid mechanics was apparently due to Lord Kelvin in a paper published in 1887 (Thomson, 1887; Schmitt, 2007a). There are several definitions of turbulence. A generalist definition would indicate "instability or irregular movements in a fluid." In physics the definition uses the dimensionless Reynolds number $Re = UL/\nu$, where U is a typical velocity scale, L a typical length scale, and ν the kinematic viscosity of the fluid. When Re is small, the flow is laminar, and the fluid motion is regular. There is a transition value where instabilities occur, and when Re is very large. Here the flow is said

2.1 Introduction

to be in a state of fully developed turbulence. Such flow is also chaotic due to the nonlinear nature of Navier-Stokes equations. Another characteristic of turbulence is its very efficient mixing property. Tennekes and Lumley (1972) insisted on the three dimensional nature of turbulence, and also on the fact that it is a dissipative system, needing, to be in a stationary state, a continuous input of energy. Other characteristics are: a system with a large number of degrees of freedom, and possessing variability on a large range of scales.

Newtonian fluid mechanics turbulence obeys the Navier-Stokes equations, which are three-dimensional, nonlinear, and deterministic. The fact that these equations are deterministic means that there are no stochastic terms in the original equations; in a given flow situation, there is a unique solution, and if the present is fully known the future is fully determined. However, there are so many degrees of freedom that this is only a theoretical consideration; since the system is chaotic, a small change in initial conditions will give rise to two extremely different solutions (Manneville, 2004; Nicolis and Nicolis, 2012). Hence, in the last few decades, physicists have been considering stochastic approaches to tackle turbulence.

Turbulence is often said to be the last question of classical physics (nineteenth century physics, before relativity and quantum mechanics) which has still not been solved today. Around the world, numerous researchers are currently working in this field: most of them in engineering or applications. There are various ways to approach turbulence: these are listed below.

- Mathematical approach: Navier-Stokes equations have been known for almost two centuries, but a full mathematical solution is out of reach for the moment. Recently, the seven Millennium Problems published in 2000 by the Clay Mathematics Institute promised one million dollars for the solution of each problem. Currently, the prize is still waiting to be attributed. A full mathematical solution would bring revolution to many fields and industries.
- Numerical simulations: Since a mathematical solution is not reachable for the moment, there are computer simulations called "Direct Numerical Simulations" (DNS). These correspond to numerically solving the Navier-Stokes equations using a discretization in time and in space. The grid size must be much smaller than the Kolmogorov scale, and the domain size of the order of the injection large scale. This is a very useful and efficient approach, which is providing more and more knowledge of many properties of turbulence. However, reachable Reynolds numbers are still too small. By noting L, the outer scale of the flow, and ℓ_0 the grid scale (hence the smallest scale of the simulation), the scale ratio is known to be of the order $L/\ell_0 \approx Re^{3/4}$. Since this is a 3D problem and the modelers want also to simulate the flow during several time steps, and enough time steps to have a large scale mixing time, the full calculation time T will be proportional to the power 4

of the scale ratio, hence $T \approx Re^3$. Any factor of 10 in the increase of the Reynolds number needs an increase of a factor of 1000 for computer calculation speeds. The Reynolds numbers for environmental situations (atmospheric or oceanic flows) are so big (10^6 to 10^9) that DNS will only likely be able to simulate such flows from the years 2030 to 2050 if the power progress of computers continues to increase exponentially in the next decades.

- Closure models of the type k-ϵ, Reynolds-average Navier-Stokes equations (RANS), or Large-Eddy Simulations (LES) families of models: For industrial applications (e.g., the flow around an airplane, around a car, or inside a turbine) engineers can estimate some statistical quantities: mean velocities and second moments (kinetic energy and stresses). These models use an average of Navier-Stokes equations on a volume using a grid size much larger than the smallest scales of turbulence. Within this approach, calculation times are much shorter. However, turbulence is a nonlinear problem: the dynamics of the average (the model) is not the same as the average of the dynamics (the averaged "reality"). The averaging of Navier-Stokes equations, needs to estimate the Reynolds stress tensor, a new term occurring because of the nonlinearity of the equations. In order to have a closed form of the averaged equations, a so-called closure must be introduced. The closure expresses the flux term (Reynolds stress tensor) as function of averages quantities. There is no exact, mathematical closure of the Navier-Stokes equations, and all closures used in models in the engineering or environmental applications are *models*, hence, false and not even approximations of reality (see a discussion in Schmitt [2007a]). There are many models proposed for this (for a review see Wilcox et al. [1998]; Piquet [1999]; Pope [2000]; Bernard and Wallace [2002]). These models provide deterministic solutions for the averaged fields in 3D, with a high prize: since there is no mathematical solution for the closure problem, the models proposed are all *ad hoc*. They assume a separation of scales between the characteristic scale of the mean velocity variations and the turbulent Lagrangian free path whose mean value is the turbulent mixing length. This separation of scales hypothesis is known for a long time to be untrue (Lumley, 1970; Hinze et al., 1974; Corrsin, 1975; Schmitt, 2007b). New turbulence models are needed that take into account non-local effects (Egolf and Weiss, 1998; Schmitt et al., 2010; Qiu et al., 2011).

- Phenomenological approach: This approach follows the path proposed originally by Kolmogorov in 1941. The idea is to look for stochastic solutions of Navier-Stokes equations (for velocity or passive scalars), using experimental properties, as well as some symmetry properties of the equations. The idea is to model the intermittency of such solutions, using appropriate stochastic models, built to reproduce the basic experimental properties. Such intermittent solutions are not proved to be mathematical solutions of the original equations. These solutions are

2.2 Richardson-Kolmogorov cascade and K41 relation

not tensorial and for most models, only unidimensional. However, they address one of the fundamental properties of turbulence, which is its production of huge fluctuations having a specific structure, a phenomenon called "intermittency." Our approach here belongs to this family.

2.2 Richardson-Kolmogorov cascade and K41 relation

We consider here a Newtonian fluid obeying the 3D incompressible Navier-Stokes equations:

$$\frac{\partial u_i}{\partial t} + u_j \frac{\partial u_i}{\partial x_j} = -\frac{\partial p}{\partial x_i} + \nu \nabla^2 u_i \qquad (2.1)$$

and the continuity equation:

$$\frac{\partial u_i}{\partial x_i} = 0 \qquad (2.2)$$

where we note $\vec{u} = (u_i)$ the velocity vector, p the pressure, and ν the kinematic viscosity of the fluid. We consider a homogeneous and isotropic situation, and further assume that the Reynolds number $Re = UL/\nu$ is large.

In such fully developed turbulent situations, Richardson, in his 1922 book, *Weather prediction by numerical process,* noticed that when drawing a rising cumulus, "the details change before the sketch can be completed." He then had the intuition that large eddies have a larger time scale than small eddies. He formulates this in the next sentence, which is now very famous: "We realize thus that: big whirls have little whirls that feed on their velocity, and little whirls have lesser whirls and so on to viscosity - in the molecular sense" (Richardson, 1922, p. 66). In modern terms, the idea here is to indicate that turbulence is a dissipative system, in which motion stops if energy is not injected. The energy is dissipated into heat at small scales, and is injected in the system at large scales. From large to small scales, there is a cascade of energy, now often called the Richardson-Kolmogorov cascade (Figure 2.1). Kolmogorov was associated to this idea since the description of Richardson was at the basis of the famous Kolmogorov relation proposed in 1941 (Kolmogorov, 1941b).

Kolmogorov assumes a local isotropy hypothesis (small-scale turbulence is homogeneous and its statistics are isotropic) for the inertial range, corresponding to scales larger than the dissipation scales and smaller than the injection scale. He assumes that velocity fluctuations between two points separated by a distance ℓ only depend on this scale and on the dissipation ϵ. By dimensional analysis the velocity fluctuations can be inferred. Let us introduce the structure function:

$$S_2(\ell) = \langle \Delta V_\ell^2 \rangle = \langle (V(M) - V(M'))^2 \rangle \qquad (2.3)$$

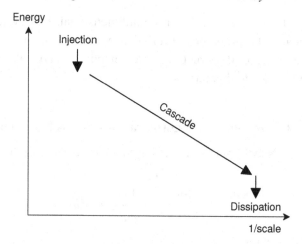

Figure 2.1 Illustration of the Richardson-Kolmogorov cascade from large to small scales.

where $\langle . \rangle$ denotes statistical average and M and M' are two points separated by a distance $d(M, M') = \ell$ belonging to the inertial range. Kolmogorov obtains the following relation, now denoted "K41" relation:

$$S_2(\ell) = C\epsilon^{2/3}\ell^{2/3} \qquad (2.4)$$

where C is a constant.

In another paper published in 1941 (Kolmogorov, 1941a), Kolmogorov obtained an exact law from the Navier-Stokes equations for the third-order moment of the longitudinal fluctuations:

$$\langle \Delta V_\ell^3 \rangle = -\frac{4}{5}\epsilon\ell \qquad (2.5)$$

where the moment is estimated here without absolute values.

Simultaneously to this physical space study, Obukhov, student of Kolmogorov, proposed a similar law in Fourier space (Obukhov, 1941):

$$E_v(k) = C'\epsilon^{2/3}k^{-5/3} \qquad (2.6)$$

where $E_v(k)$ is the Fourier power spectrum of the velocity, C' is a constant, and k the wavenumber. This law has been verified in many studies, in the atmosphere (Gurvich, 1960), the ocean (Grant et al., 1962), and in the laboratory (Champagne, 1978; Gagne, 1980; Anselmet et al., 1984). It has become universally used in describing fully developed turbulence (see an example in Figure 2.2).

K41 relation can be used to statistically express the Reynolds number using the scale ratio. In fact, for a given scale ℓ belonging to the inertial range, let us consider

2.2 Richardson-Kolmogorov cascade and K41 relation

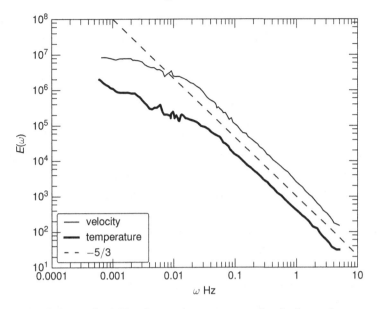

Figure 2.2 (Color online) Fourier power spectrum of velocity and temperature time series recorded in the atmosphere, displaying 5/3 power law decay over about two decades.

ΔV_ℓ the velocity fluctuations at this scale. A local Reynolds number corresponding to this scale can be defined as:

$$Re_\ell = \frac{\Delta V_\ell \ell}{\nu} \quad (2.7)$$

Let us introduce the scale η, also called the Kolmogorov scale, for which $Re_\eta \approx 1$. We have $\Delta V_\eta \eta = \nu$ and using the Kolmogorov relation $\Delta V_\ell \approx \epsilon^{1/3} \ell^{1/3}$ for $\ell = \eta$ we have $\epsilon^{1/3} \eta^{4/3} = \nu$, hence the Kolmogorov scale η is given by:

$$\eta = \left(\frac{\nu^3}{\epsilon}\right)^{1/4} \quad (2.8)$$

This gives the smallest scale of turbulent motion using the viscosity, a parameter of the fluid, and the dissipation, a parameter of the flow. The Kolmogorov's scale separates the turbulent scales and the viscous scales: for larger scales we are in the inertial range corresponding to the turbulence scales; below this scale we are in the viscous range where viscosity begins to be important. Using typical values of the dissipation and viscosity in the atmosphere and ocean, it is usually found that Kolmogorov's scale is of the order of millimeters, both in the atmosphere and ocean.

In the general framework of the Richardson-Kolmogorov cascade, energy is thus injected at large scales (of the order of meters, kilometers, or hundreds of kilometers

2.3 Intermittency and Yaglom's cascade model

2.3.1 Experimental results and Kolmogorov's refined similarity hypothesis

Before the experimental verification of Kolmogorov's scaling power spectrum in the 1960s, some experimental measurements in a wind tunnel were carried out at Cambridge a few years after World War II. Batchelor and Townsend (1949) experimentally estimated a proxy of the small scale dissipation using hot wires. They found that "[t]he basic observation which requires explanation is that activation of large wave-numbers is very unevenly distributed over space," and "[a]s the wavenumber is increased the fluctuations seem to tend to an approximate on-off, or intermittent, variation." (p. 252). It was the first introduction of the word intermittency in turbulence. In their conclusion, the authors indicated that they had no explanation of such phenomenon, but that apparently, the "rhyme originally due to L. F. Richardson (...) must be taken more literally in turbulence than had been thought." A modern example of a proxy of the dissipation field is shown in Figure 2.3.

A few years later, at two famous meetings on turbulence held in August and September 1961 in Marseilles, Obukhov and Kolmogorov proposed a new model to take into account such intermittency in turbulent fluctuations. They proposed to introduce a new quantity, the averaged dissipation over volume of typical scale ℓ, which we denote here Π_ℓ:

$$\Pi_\ell(x) = \int_{B_\ell(x)} \epsilon(x')dx' \tag{2.9}$$

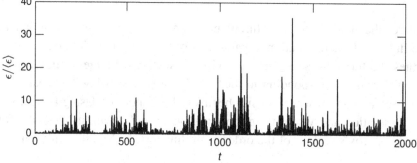

Figure 2.3 A modern example of a proxy of the dissipation field recorded in the atmosphere. The series is normalized by its mean, showing the large fluctuations.

2.3 Intermittency and Yaglom's cascade model

where $B_\ell(x)$ is the volume centered at position x and of diameter ℓ. The dissipation ϵ is here a well-defined small-scale quantity, having locally large fluctuations. In this framework the local average was considered to be the relevant quantity to study and modelize. Then Obukhov assumed that this local average followed a lognormal probability density function. Kolmogorov credits Obukhov for this idea and adopts the same hypothesis. Kolmogorov (1962) does not give any justification to this hypothesis; he only indicates "It is natural to suppose that ..." (Kolmogorov, 1962, p. 83). On the other hand, Obukhov (1962) is more precise, since he indicates that a distribution of a positive quantity can always be approached by a lognormal distribution with the correct value for the first two moments (Obukhov, 1962, p. 79).

The model proposed by these two researchers is a lognormal probability density function (pdf) for the averaged dissipation Π_ℓ, with a variance σ_ℓ^2 of $\log \Pi_\ell$ assumed to be given by (Kolmogorov, 1962):

$$\sigma_\ell^2 = A + \mu \log L/\ell \tag{2.10}$$

where A is a constant depending on the macrostructure of the flow, μ is a universal constant and L is the external scale.

This can be written on the following form, corresponding to what is now called Refined Similary Hypothesis (RSH):

$$\Delta V_\ell \stackrel{d}{=} \Pi_\ell^{1/3} \ell^{1/3} \tag{2.11}$$

where "$\stackrel{d}{=}$" means equality in distribution. Here we consider the structure functions $S_q(\ell) = \langle |\Delta V_\ell|^q \rangle$ for moments of order $q > 0$ of the absolute value of the fluctuations (the absolute value here is proposed to be able to consider also non-integer moments q). This equation statistically relates the velocity fluctuations at a given scale, with the averaged dissipation, using a rescaling factor (the power 1/3 of the scale). It is hence called a "scaling" relation. In this framework the structure functions are scaling of the form:

$$S_q(\ell) \approx \ell^{\zeta(q)} \tag{2.12}$$

where "$\approx$" means here proportionality within slowly varying constants, that may depend on q, and the exponent $\zeta(q)$ is the scaling exponent of the structure functions. A fixed point is known from the 4/5 Kolmogorov's relation,

$$\zeta(3) = 1. \tag{2.13}$$

Taking into account this fixed point, the structure functions according to Kolmogorov (1962) are given by a quadratic expression:

$$\zeta(q) = \frac{q}{3} - \frac{\mu}{18} q(q-3) \tag{2.14}$$

We verify that $\zeta(0) = 0$, $\zeta(3) = 1$ and $\zeta(q)$ is a convex function. The corresponding Fourier power spectrum is scaling with a slope given by:

$$\beta = 1 + \zeta(2) = \frac{5}{3} + \frac{\mu}{9} \qquad (2.15)$$

which corresponds to a small correction, since μ can be experimentally estimated to be $\mu = 0.25 \pm 0.05$ (Sreenivasan and Kailasnath, 1993). With this correction the slope is about 1.69.

2.3.2 Power-law correlations of the dissipation and Yaglom's multiplicative cascade model

A few years after Kolmogorov's 1962 intermittency model proposal, several researchers studied the dynamical structure of the dissipation field. By differentiating velocity time series, Gurvich and Zubkovskii (1963) estimated the Fourier power spectrum of the dissipation and found that it obeys a power-law:

$$E_\epsilon(k) \approx k^{-\theta} \qquad (2.16)$$

with $\theta \simeq 0.6$. This corresponds to a field possessing long-range power-law correlations:

$$\langle \epsilon(x)\epsilon(x+r) \rangle \approx r^{-(1-\theta)} \qquad (2.17)$$

These results were confirmed by other experimental studies in the atmosphere, such as Pond and Stewart (1965). Such long-range power-law correlation is a specific dynamical property, which was not explicitly contained in Kolmogorov and Obukhov's 1962 model.

There were thus two proposals for the dissipation ϵ: on the one hand a theoretical proposal by Obukhov and Kolmogorov of lognormal statistics of averaged dissipation, and on the other hand experimental results showing that the small scale dissipation has long-range power-law correlations. This lead Yaglom (1966) to propose a multiplicative cascade model (see also Gurvich and Yaglom [1967]) that would be compatible with both proposals. This model was innovative because it put forward a mechanism to generate what was only assumed by Kolmogorov (1962). It is currently, 50 years after its introduction, still the basis of all discrete multiplicative multifractal models.

The Yaglom multiplicative model assumes a lognormal pdf, but it is now known that this condition is not needed and the model is more general. This model is discrete in scale; we will see in the next section how to densify it to have continuous models. This model is multiplicative and recursive. The multiplicative hypothesis generates huge fluctuations, and the recursivity in scale generates long-range

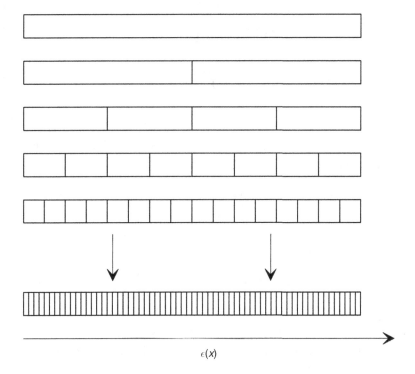

Figure 2.4 A schematic representation of a discrete multiplicative cascade.

correlations, which give spatially or temporally these large fluctuations, and their intermittent character.

The eddy cascade is represented by cells; each cell is associated with a random variable. All these random variables obey the same law, and are positive and independent for each cell. The external scale corresponds to a unique cell of size L. The model is discrete in scale, and the next step corresponds to two cells, each of size $L/2$. The cascade is iterated and at step p there are 2^p cells, each of size $L/2^p$, until the dissipation scale is reached, $\ell_0 = L/2^n$ (see Figure 2.4). To reach the small scale dissipation, the energy goes through all intermediary scales to mimic the local interaction in the energy cascade. For all points, dissipation can be written as the product of n random variables:

$$\epsilon(x) = \prod_{i=1}^{n} W_{i,x} \tag{2.18}$$

where $W_{i,x}$ is the random variable corresponding to position x and level i in the cascade. Since, by hypothesis, for two different levels, random variables are independent, the statistical moments of ϵ can be written:

$$\langle(\epsilon(x))^q\rangle = \langle\left(\prod_{i=1}^{n} W_{i,x}\right)^q\rangle \tag{2.19}$$

$$= \prod_{i=1}^{n}\langle(W_{i,x})^q\rangle \tag{2.20}$$

$$= \langle W^q\rangle^n \tag{2.21}$$

Introducing the scale ratio $\lambda = L/\ell_0 = 2^n$ between the injection scale and the smallest scale, this gives the scaling law for the statistical moments:

$$\langle(\epsilon(x))^q\rangle \approx \lambda^{K(q)} \tag{2.22}$$

with

$$K(q) = \log_2\langle W^q\rangle \tag{2.23}$$

Equation (2.22) is a generic property of multifractal fields obtained through a multiplicative cascade. For a discrete cascade, the only constraint on the random variable W is to be positive. The form of Equation (2.23) shows that, for non-zero random variables, $K(q)$ is (to a factor $\log 2$) the second Laplace characteristic function of the random variable $\log W$, which shows that it is a convex function (see Feller [1971]).

Conversely, the autocorrelation of ϵ can be estimated using an approach originally proposed by Yaglom (1966): the correlation $\langle\epsilon(x)\epsilon(x+r)\rangle$ can be decomposed in a product of random variables. A precise calculation needs to estimate at which step the two ancestors of each variable situated at a distance r one from each other were different (see Figure 2.5). In fact, the distance r corresponds, on average, to p steps, where $2^p \approx r$, and it is possible to state that the first $n-p$ steps were common to each path (and hence correspond to identical random variables) and there was a path "bifurcation" for the last p steps (hence corresponding to independent random variables). This can be formalized in the following way:

$$\langle\epsilon(x)\epsilon(x+r)\rangle = \langle\prod_{i=1}^{n} W_{i,x} \prod_{j=1}^{n} W_{j,x}\rangle \tag{2.24}$$

$$= \prod_{i=n-p}^{n}\langle W_{i,x}\rangle \prod_{j=n-p}^{n}\langle W_{j,x}\rangle \prod_{i=1}^{n-p}\langle W_{j,x}^2\rangle \tag{2.25}$$

$$= \langle W\rangle^{2p}\langle W^2\rangle^{n-p} \tag{2.26}$$

By using the relations $p\log 2 \approx \log r$ and $n\log 2 = \log\lambda$, and equation (2.16), we obtain:

$$\langle\epsilon(x)\epsilon(x+r)\rangle \approx \lambda^{K(2)} r^{2K(1)-K(2)} \tag{2.27}$$

2.4 Multifractal properties of multiplicative cascades

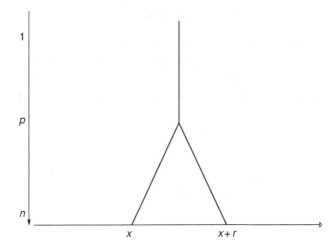

Figure 2.5 Mechanism generating long-range power-law correlations: two points separated by a distance r have a common ancestor at step p such that $2^p \approx r$.

for a given cascade, λ is a constant and this relation expresses, for r belonging to the inertial range (scales between ℓ_0 and L), the fact that small-scale dissipation is a stochastic process having long-range power-law correlations.

We finally obtain the autocorrelation of ϵ:

$$\langle \epsilon(x)\epsilon(x+r) \rangle \approx r^{-\mu} \qquad (2.28)$$

where we have introduced the intermittency exponent $\mu = K(2)$. In the original paper, Yaglom (1966) chooses for W a lognormal law, giving a quadratic form $K(q) = Bq^2 - (2B - \mu/2)q$. Using $K(1) = 0$ we obtain:

$$K(q) = \frac{\mu}{2}(q^2 - q) \qquad (2.29)$$

But this choice is not the only one, especially for discrete models. This is discussed in the next section.

2.4 Multifractal properties of multiplicative cascades

We consider here the multiplicative cascade model presented in the previous section, with a random variable W characterized by its second Laplace characteristic function $\Psi(q) = \log\langle W^q \rangle$, which is assumed to exist, and have a general form (not necessary lognormal). We shall then explore the multifractal properties of this cascade.

2.4.1 Singularities

Multiplicative cascade models generate an intermittent and positive field, with values becoming more and more intermittent and with larger fluctuations, as the scale ratio increases. This corresponds to so-called singularities with multifractal properties. We shall consider a multiplicative cascade developed over a total scale ratio of

$$\lambda = L/\ell \gg 1 \tag{2.30}$$

where $\ell \ll 1$ is the smallest scale of the cascade and L the fixed outer scale. Since the scale ratio is large, the field ϵ_ℓ which is generated at scale ℓ becomes singular, characterized by singularities γ defined by:

$$\epsilon_\ell \approx \ell^{-\gamma} \tag{2.31}$$

These singularities can be characterized using their probability distribution involving the codimension function $c(\gamma)$ (Schertzer and Lovejoy, 1987; Frisch and Matsumoto, 2002):

$$\Pr(\gamma > \gamma') \approx \ell^{c(\gamma')} \tag{2.32}$$

Each γ value has a different fractal dimension; the infinite number of dimension is at the origin of the word "multifractal" for such an object (Grassberger and Procaccia, 1983; Hentschel and Procaccia, 1983; Parisi and Frisch, 1985). The multifractal framework was in fact developed in parallel in both the field of chaos theory and strange attractors, and in the field of turbulence. Grassberger and Procaccia (1983) and Hentschel and Procaccia (1983) introduced the idea of infinite number of generalized dimensions for the singular measures of strange attractors, and Parisi and Frisch (1985) coined the word "multifractal" in the framework of intermittent turbulence.

The fractal dimension $d(\gamma)$ is expressed using the number $N(\gamma)$:

$$N(\gamma) \approx \lambda^{d(\gamma)} \tag{2.33}$$

obtained using a box-counting (Mandelbrot, 1983). By normalizing by the total number of events $N_T \approx \lambda^D$, where D is the dimension of the space ($D = 1$ for a time series), one obtains the probability. Hence the relation:

$$c(\gamma) = D - d(\gamma) \tag{2.34}$$

The codimension function is positive, increasing, and convex (see an example in Figure 2.6).

The notation $\alpha = 1 - \gamma$ and $f(\alpha) = d(\gamma)$ is often used in the multifractal literature (for $D = 1$). In the data analysis chapters, we will display the singularity spectrum by considering the $f(\alpha)$ curve.

2.4 Multifractal properties of multiplicative cascades

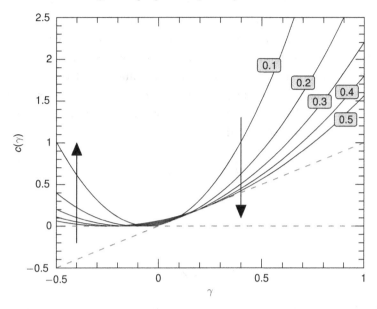

Figure 2.6 Schematic graphical representation of the codimension function, which is increasing and convex. Here the plot is shown for the lognormal case, with different values of μ.

2.4.2 Legendre transform

To characterize the statistics of the multifractal field ϵ_ℓ, there is the probability density of singularities, or the moment function

$$\langle (\epsilon_\ell)^q \rangle = \int \ell^{-q\gamma} dp(\gamma) \approx \ell^{\min_\gamma \{c(\gamma) - q\gamma\}} \tag{2.35}$$

This corresponds to a scaling relation for the statistical moments:

$$\langle (\epsilon_\ell)^q \rangle \approx \ell^{-K(q)} \tag{2.36}$$

where the moment function $K(q)$ and the codimension function $c(\gamma)$ are related by a Legendre transform (Parisi and Frisch, 1985; Halsey et al., 1986):

$$K(q) = \max_\gamma \{q\gamma - c(\gamma)\} \tag{2.37}$$

Parisi and Frisch (1985) introduced the Legendre transform for structure functions scaling exponents (see Equation 2.38), and Halsey et al. (1986) did it later in the framework of strange attractors. Mandelbrot (1991) showed that such a relation can also be obtained in the framework of large deviations theory. See also the discussion in Collet and Koukiou (1992) and in Frisch (1995).

Such a relation expresses a one-to-one relationship between the statistical moment and the singularity: each singularity γ is associated with a moment q and reciprocally. Large moments correspond to large singularities and extreme events. Low order moments are associated to medium singularities. In particular, for moment of order $q = 1$, characterized by $K(1) = 0$, there is a singularity $\gamma_1 = C_1$ such as $c(C_1) = C_1$. When $K(q)$ and $c(\gamma)$ can be differentiated, the Legendre transform can also be written as:

$$\begin{cases} q = c'(\gamma) \\ K(q) = q\gamma - c(\gamma) \end{cases} \quad (2.38)$$

or:

$$\begin{cases} \gamma = K'(q) \\ c(\gamma) = q\gamma - K(q) \end{cases} \quad (2.39)$$

As an illustration of the use of the latter relations, we see here the Legendre transform for the lognormal model, as the prototype of multifractal models. There is a quadratic expression for $K(q)$, given by Equation (2.29). The Legendre transform on the form of Equation (2.38) first give:

$$\gamma = K'(q) = \mu q - \frac{\mu}{2} \quad (2.40)$$

Then:

$$c(\gamma) = qK'(q) - K(q) = \frac{\mu}{2}q^2 \quad (2.41)$$

giving finally, using Equation (2.40), the quadratic expression:

$$c(\gamma) = \frac{\mu}{2}\left(\frac{\gamma}{\mu} + \frac{1}{2}\right)^2 = \frac{1}{4C_1}(\gamma + C_1)^2 \quad (2.42)$$

We see that here $C_1 = \mu/2$ and $c(C_1) = C_1$.

We recognize also that $K(q) = \Psi(q)/\log 2$, hence $K(q)$ is, to a multiplicative factor, a cumulant generating function, or a second Laplace characteristic function. It is therefore convex, and its Legendre transform, $c(\gamma)$, is also convex. The Legendre transform between $K(q)$ and $c(\gamma)$ is also related to the Legendre transforming existing, in the large deviation theory, between the rate function (or entropy function, linked with the probability distribution) and the cumulant generating function (Mandelbrot, 1991; Frisch, 1995).

2.4.3 Dressed and bare properties of multiplicative cascades

Let us also mention the difference between " bare" and "dressed" multiplicative cascades. In the previous section we considered the scaling properties of a cascade developing from a large scale to a small scale $\ell \ll 1$: this is the bare cascade, following Mandelbrot's (Mandelbrot, 1974) and Schertzer and Lovejoy's (Schertzer and Lovejoy, 1987, 1992) terminologies. A dressed quantity is the same cascade, developed over a very small scale ℓ_0 (e.g., the dissipation scale η) and then averaged over a ball B_ℓ of size $\ell > \ell_0$. This difference is important because bare cascades can theoretically develop over a very large scale ratio, leading to huge local fluctuations, whereas dressed quantities, on the contrary, correspond to observables at a given finite scale ℓ (which is often much larger than the small scale ℓ_0).

The difference is hence of practical importance since applied scientists often analyze experimental data by "dressing" a small scale intermittent field, i.e., by averaging the data over balls of size ℓ, and considering the statistics of the result versus ℓ (Meneveau and Sreenivasan, 1987; Schmitt et al., 1992; de Montera et al., 2009). In other words, the bare cascade is not observable: it is a hidden quantity; observables are dressed fields. The bare cascade develops from large to small scales. Observables are analyzed by doing data analysis, going from small to large scales, by coarse-graining the high resolution measurements.

Let us introduce the following quantity, with the same relations as in Equation (2.9):

$$\Pi_\ell(x) = \int_{B_\ell(x)} \epsilon_{\ell_0}(x')dx' \qquad (2.43)$$

as the dressed quantity, which is the average of the cascade ϵ_{ℓ_0} (developed over a large scale ratio $\Lambda = L/\ell_0 \gg 1$) over a ball $B_\ell(x)$ of size ℓ centered in x. It has been shown (Mandelbrot, 1974; Schertzer and Lovejoy, 1987) that the dressed quantity is also scaling with the following property:

$$\langle (\Pi_\ell)^q \rangle \sim \ell^{-K_D(q)} \qquad (2.44)$$

where $K_D(q)$ is the dressed moment function. This verifies:

$$K_D(q) = \begin{cases} K(q); & q \le q_D \\ \gamma_m q - c(\gamma_m); & q > q_D \end{cases} \qquad (2.45)$$

where γ_m is a maximal singularity and q_D is a critical order of moment given by (Mandelbrot, 1974; Kahane, 1985; Schertzer and Lovejoy, 1987):

$$K(q_D) = (q_D - 1)D \qquad (2.46)$$

In probability theory, this is called "multiplicative chaos" and corresponds in itself to a new family of random processes (Kahane, 1985; Robert and Vargas, 2010; Rhodes and Vargas, 2014), which are not developed here.

Equation (2.45) shows that for $q > q_D$, K_D is linear (Schertzer and Lovejoy, 1992; Schmitt et al., 1994; Lashermes et al., 2004; Muzy et al., 2006) with a slope, which increases with the number of realizations (see Schmitt et al. [1994] for an experimental analysis); it corresponds to a property of divergence of moments. Due to the analogy with statistical mechanics, such change of behavior of the dressed field for this critical moment has been called a "multifractal phase transition" (Schertzer and Lovejoy, 1992).

It must be also noted that, even without such divergence of moments, the dressed moment function $K_D(q)$ behaves linearly for large values of q, due to sampling limitations. Indeed, in case of finite sampling, a maximum singularity γ_s is found in the data, associated with a critical moment $q_s = c'(\gamma_s)$. For moments $q \geq q_s$, $K_D(q)$ will be linear, with behavior (Schertzer and Lovejoy, 1992):

$$K_D(q) = q\gamma_s - c(\gamma_s) = \gamma_s(q - q_s) + K(q_s) \qquad (2.47)$$

The main difference with the previous situation (Equation [2.45]) is the fact that, when the behavior is linear due to of sampling limitation, the critical moment is increasing with the sampling, and the linear asymptotic form of $K_D(q)$ is the tangent of $K(q)$ at $q = q_s$.

2.4.4 Discrete cascade models

We consider in the following several generic discrete multiplicative cascades following Yaglom's model. Each discrete cascade model is mainly characterized by the scaling moment function $K(q)$.

Black-and-white model: the β-model

One of the simplest cascade models is a "black-and-white" model generating a linear function $K(q)$, hence a mono-fractal model: the β-model, introduced under this name by Frisch et al. (1978), but already presented in Mandelbrot (1974) using an approach originally proposed by Novikov and Stewart (1964). In fact, Novikov and Stewart's model, which they characterized as "pulses into pulses," is not a cascade model, it is instead a hypothesis on averaged quantities, without explaining how to build a process possessing such property. In other words, they described the dressed properties without explicitly providing a mechanism to produce bare cascades.

According to the β model, at each step of the cascade, the random variable W can take only two values, 0 or β, with $0 < \beta < 1$:

2.4 Multifractal properties of multiplicative cascades

$$\begin{cases} \Pr(W = 0) = 1 - \beta \\ \Pr(W = 1/\beta) = \beta \end{cases} \quad (2.48)$$

This model verifies $K(1) = 0$, and the moment function $K(q)$ is given by:

$$K(q) = C_1(q - 1) \quad (2.49)$$

where the parameter $0 < C_1 < 1$ is given by $C_1 = -\log_2 \beta$.

This model produces a linear moment function. It is hence mono-fractal, with a unique fractal dimension $d_s = D + \log_2 \beta = D - C_1$.

α-model and p-model

Other discrete cascade models have been proposed, such as the α-model (Schertzer and Lovejoy, 1984), which is a binomial model having two possible values for W, $W+ = 2^{\gamma+}$ and $W- = 2^{\gamma-}$:

$$\begin{cases} \Pr(W = 2^{\gamma+}) = 2^{-c} \\ \Pr(W = 2^{\gamma-}) = 1 - 2^{-c} \end{cases} \quad (2.50)$$

The coefficients $\gamma+$, $\gamma-$ and c are linked by the normalization condition $\langle W \rangle = 1$. The corresponding moment function writes:

$$K(q) = \log_2 \left(2^{q(\gamma+)-c} + (1 - 2^{-c})2^{-q(\gamma-)} \right) \quad (2.51)$$

A more restrictive version of this model (taking $c = 1$) was put forward by Meneveau and Sreenivasan (1987), with a new name, the p-model:

$$\begin{cases} \Pr(W = 2p) = 1/2 \\ \Pr(W = 2 - 2p) = 1/2 \end{cases} \quad (2.52)$$

This gives a conservative model ($\langle W \rangle = 1$), and the moment function writes:

$$K(q) = q - 1 + \log_2 (p^q + (1 - p)^q) \quad (2.53)$$

with $0 < p < 1$.

Random β-model

Another discrete cascade model was put forward by Benzi et al. (1984), the random β-model, by taking a random parameter β in the β-model. This gives the scaling moment function:

$$K(q) = \Psi_0(1 - q)/\log 2 \quad (2.54)$$

where $\Psi_0(q) = \log \langle \beta^q \rangle$ is the second characteristic function of the random variable β. The authors chose for β the following binomial distribution:

$$\begin{cases} \Pr(\beta = 1/2) = x \\ \Pr(\beta = 1) = 1 - x \end{cases} \quad (2.55)$$

where $0 < x < 1$ is a parameter, giving finally (the authors have chosen $x = 0.125$):

$$K(q) = \log_2\left(1 - x + x2^{q-1}\right) \tag{2.56}$$

2.4.5 Continuous cascades and infinitely divisible models

Densification: continuous models

Discrete models are not realistic since they involve an elementary scale ratio (often a scale ratio of 2) and hence the output is not translationally invariant. A discrete cascade becomes realistic through densification. This is done by considering as constant the larger scale ratio $\Lambda \gg 1$, and increasing the number of elementary steps. For this, we consider the discrete cascades presented above, and replace the elementary scale ratio, which was 2, by a variable parameter $\lambda_1 > 1$. The total scale ratio is still:

$$\Lambda = \lambda_1^n \tag{2.57}$$

which is large but fixed. In this case, to densify the cascade corresponds to increase n so that the elementary scale ratio decreases:

$$\lambda_1 = \Lambda^{1/n} \to 1^+ \tag{2.58}$$

The multiplicative structure of the cascade process is written as follows, for any value of n:

$$\log \epsilon(x) = \sum_{i=1}^{n} \log W_{i,x} \tag{2.59}$$

This corresponds to the definition of an infinitely divisible (ID) random variable (see Feller [1971]): $\log \epsilon$ has an infinitely divisible probability density. Hence continuous cascade models correspond to log-ID probability distributions (Novikov, 1994). This considerably restricts the class of cascade processes, since ID laws are a specific type of random variables. This class of models is still infinite, since there is an infinite number of ID laws.

Novikov (Novikov, 1969, 1971) previously mentioned the concept of scale similarity, corresponding to a continuity in scales, and he obtained scale invariant laws for moments, using a function analogous to our $K(q)$. To obtain these results, he did not introduce a cascade and used "breakdown coefficients" which are dressed quantities. Schertzer and Lovejoy (1987) introduced the concept of "continuous cascades," using the Lévy-Khintchine formula for the characteristic function of ID variables. In the rest of their paper, they focus on stable laws, a specific subset of ID variables. Later, Novikov (1994) showed the link between continuous cascades and log-ID models.

2.4 Multifractal properties of multiplicative cascades

We have mentioned below two important classes of Log-ID models: the log-stable (also called "universal multifractals"; Schertzer and Lovejoy, 1987; Kida, 1991; Schmitt et al., 1992; Schertzer et al., 1997) and the log-Poisson model (She and Lévêque, 1994; Dubrulle, 1994; She and Waymire, 1995).

Log-stable model

The Lévy stable laws (Lévy, 1937) are important laws belonging to the ID family: in addition to being decomposable (ID property) they are also stable and attractive under addition. These laws are interesting for multiplicative cascades since the power $a > 0$ of a log-stable process is still log-stable, which is not the case for other multifractal models for $a \neq 1$. The most important parameter of a Lévy-stable random variable is the parameter $0 \leq \alpha \leq 2$. The Gaussian law is stable and corresponds to the value $\alpha = 2$. For $\alpha < 2$, these laws are sometimes called "wild" since the moment of order α or larger are diverging. For more information on stable laws, see Gnedenko and Kolmogorov (1954); Zolotarev (1986); Samorodnitsky and Taqqu (1994); Janicki and Weron (1994); Nikias and Shao (1995); Uchaikin and Zolotarev (1999).

Stable laws are also characterized by a parameter β which corresponds to the asymmetry between positive and negative values. In the case of log-stable laws, since one must consider moments of order $q > 0$ of the exponential of such laws, one must take the parameter $\beta = -1$, corresponding to hyperbolic "wild" fluctuations only for negative values (Schertzer and Lovejoy, 1987; Kida, 1991). In this framework the log-stable model is characterized by:

$$K(q) = \frac{C_1}{\alpha - 1}(q^\alpha - q) \quad ; \quad c(\gamma) = C_1 \left(\frac{\gamma}{C_1 \alpha'} + \frac{1}{\alpha} \right)^{\alpha'} \quad (2.60)$$

where $C_1 = K'(1)$, and α' is such that:

$$\frac{1}{\alpha'} + \frac{1}{\alpha} = 1 \quad (2.61)$$

the quadratic lognormal model is recovered for $\alpha = 2$.

Log-Poisson model

On the other hand, one of the most characteristic ID laws is the Poisson law. This distribution is not stable, since a linear combination of Poisson variables is no more Poisson. Nevertheless, the log-Poisson model, which was put forward in the 1990s by several authors (She and Lévêque, 1994; Dubrulle, 1994; She and Waymire, 1995) has been quite popular in the turbulence community. This model corresponds to the following $K(q)$ function:

$$K(q) = c[(1 - \beta)q - 1 + \beta^q] \quad (2.62)$$

where $c > 0$, $0 < \beta < 1$. The original model (She and Lévêque, 1994) proposes $\beta = 2/3$ and $c = 2$, but other values of the parameters have been put forward: e.g., Chen and Cao (1995) chose the constraint $c(1 - \beta) = 1$, and chose $\beta = 7/9$ et $c = 9/2$.

We can see that $\gamma = K'(q) = c\left((1 - \beta) + \beta^q \log \beta\right) \leq c(1 - \beta) = \gamma_{\max}$. When $q \to \infty$, $\beta^q \to 0$ and $\gamma \to \gamma_{\max}$. There is a maximum singularity $\gamma_{\max}$ in this model and the asymptotic form of the moment function is linear $K(q) = q\gamma_{\max} - c$; the parameter c is the codimension of this maximum singularity (Schertzer et al., 1995).

There are other log-ID models, such as the log-Gamma model (Saito, 1992), or log-inverse Gaussian model (Anh et al., 2008). There is potentially an infinity of log-ID models. All are candidates for stochastic continuous cascade modeling.

2.5 Velocity fluctuations and multiscaling properties

After the exploration in the previous section of multiplicative cascade statistical properties and their multifractal characteristics, we now return to the velocity field. Fields obtained through a multiplicative cascade (e.g., typically the dissipation field in turbulence) are usually experimentally studied using coarse-graining to go from small to large scales. Other methods exist, involving, e.g., wavelets (see the next sections devoted to scaling methods). The velocity field, as well as the passive scalar field in turbulence, and also many other fields, such as financial fluctuations, fracture process, and climate fluctuations, cannot be directly modeled using multiplicative processes.

Several methods have also been published for such process, and in this section we use only the structure functions notation, since this was originally used by Kolmogorov. Velocity fluctuations have multifractal properties and the word "multifractal" is used classically for the output of multiplicative cascades, as well as for fields such as the velocity in turbulence. Such fields have been also called "non-conservative" multifractals (Pecknold et al., 1993; Schertzer et al., 1997), "multi-affine" (Barabási et al., 1991; Benzi et al., 1993b; Kuramoto et al., 1998), or in the time domain, "non-stationary" (Schertzer and Lovejoy, 1987; Davis et al., 1993, 1994). For time series, such processes belong mathematically to the set of non-stationary stochastic processes with stationary increments (Feller, 1971).

2.5.1 Relations between the velocity fluctuations and multiplicative cascades

Yaglom's multiplicative cascades, together with Kolmogorov's RSH (Equation [2.11]) to relate velocity statistics to the cascading quantity, the energy flux, are

2.5 Velocity fluctuations and multiscaling properties

now classically used to describe turbulence intermittency in the inertial range, from the injection scale to the Kolmogorov scale. We recall the RSH and the scaling relations for the velocity and the dressed energy dissipation Π_ℓ:

$$\begin{cases} \Delta V_\ell \stackrel{d}{=} \Pi_\ell^{1/3} \ell^{1/3} \\ \langle |\Delta V_\ell|^q \rangle \approx \ell^{\zeta(q)} \\ \langle (\Pi_\ell)^q \rangle \approx \ell^{-K_D(q)} \end{cases} \quad (2.63)$$

By taking moments of order q of the RSH relation, we obtain a relation between scale invariant moment functions:

$$\zeta(q) = \frac{q}{3} - K_D\left(\frac{q}{3}\right) \quad (2.64)$$

The conservation of the cascade corresponds to $K_D(1) = 0$, leading to $\zeta(3) = 1$, which is a fixed point for the velocity structure functions coming from Equation (2.3). We see also that $\mu = K_D(2)$ gives the following relation:

$$\mu = 2 - \zeta(6) \quad (2.65)$$

Furthermore, due to the intermittency effect, there may be some variations around the K41 spectral exponent 5/3. Indeed the velocity power spectrum may be written $E_v(k) \approx k^{-\beta}$ with the following relation:

$$\beta = 1 + \zeta(2) \quad (2.66)$$

and for intermittency models $\zeta(2)$ may be slightly different than 2/3, as already seen in the lognormal case.

2.5.2 Classical models for the velocity structure functions

Since Equation (2.64) relates the moment functions of dissipation and velocity structure functions, we do not need to repeat here all the cascade models for the velocity. For completeness, we have rewritten only the most classical ones, with some values of their parameters that can be found in the literature.

First, the two linear models, for the K41 non-intermittent initial proposal, and for the β-model:

$$\zeta_{K41}(q) = \frac{q}{3} \quad (2.67)$$

$$\zeta_\beta(q) = (1 - C_1)\frac{q}{3} + C_1 = \frac{q}{3}(1 + \log_2 \beta) - \log_2 \beta \quad (2.68)$$

The K41 scaling moment function is the same as for a fractional Brownian motion, with parameter $H = 1/3$ (see next section).

For other cascade models, $\zeta(q)$ is nonlinear; we provide below the three classical log-ID cascade model expressions. For the log-stable model (Schertzer and Lovejoy, 1987; Kida, 1991; Schertzer et al., 1997):

$$\zeta_{LS}(q) = \frac{q}{3}\left(1 + \frac{C_1}{\alpha - 1}\right) - \frac{C_1}{3^\alpha(\alpha - 1)}q^\alpha \tag{2.69}$$

where the parameters $0 \leq C_1 \leq 1$ and $0 \leq \alpha \leq 2$ proposed are, with the present notations, $\alpha = 1.65$ and $C_1 = 0.176$ (Kida, 1991) and $\alpha = 1.5$ and $C_1 = 0.15$ (Schmitt et al., 1996; Schertzer et al., 1997). The lognormal model (Kolmogorov, 1962; Obukhov, 1962) is recovered by taking $\alpha = 2$ and using the notation $\mu = 2C_1$:

$$\zeta_{LN}(q) = \frac{q}{3}\left(1 + \frac{\mu}{2}\right) - \frac{\mu}{18}q^2 \tag{2.70}$$

where the value $\mu = 0.25$ is classically taken (Sreenivasan and Kailasnath, 1993). And finally, for the log-Poisson cascade model (She and Lévêque, 1994; Dubrulle, 1994; She and Waymire, 1995):

$$\zeta_{LP}(q) = \frac{q}{3}(1 - c(1 - \beta)) + c\left(1 - \beta^{q/3}\right) \tag{2.71}$$

where the values of the parameters $c > 0$, $0 < \beta < 1$ were put forward originally as $\beta = 2/3$ and $c = 2$, with also the proposal of Chen and Cao (1995): $\beta = 7/9$ and $c = 9/2$.

Let us note that the log-stable model provides for $\zeta(q)$ a linear expression with a power-law nonlinear term, whereas the log-Poisson model provides a linear expression with an exponential nonlinear term. Figure 2.7 represents these five classical models for the velocity structure functions scaling exponents: the K41 model; the β model with $C_1 = 0.125$ (parameter chosen to be tangent to experimental data at $q = 3$); the lognormal model with $\mu = 0.25$; the log-Poisson model with $c = 2$ and $\beta = 2/3$, and the log-stable model with $\alpha = 1.5$ and $C_1 = 0.15$. The three log-ID models are very close to each other, until the moment of about 6.

2.5.3 Legendre transform for the velocity and large moment behavior

The Legendre transform has been expressed above for multiplicative cascades; one may characterize the velocity fluctuations ΔU_ℓ at scale ℓ through the singularities h and their codimension $c(h)$ (Parisi and Frisch, 1985; Schertzer and Lovejoy, 1987; Frisch, 1995; Schertzer et al., 1997):

$$\Delta U_\ell \sim \ell^h; \quad p(\Delta U_\ell) \sim \ell^{c(h)} \tag{2.72}$$

2.5 Velocity fluctuations and multiscaling properties

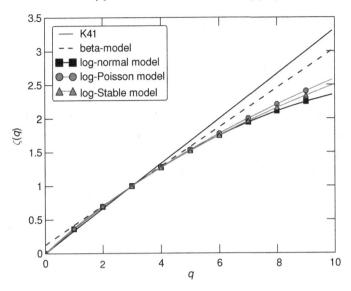

Figure 2.7 Representation for moments from 0 to 10 of five classical models for the velocity structure functions scaling exponents: the K41 model; the β model with $C_1 = 0.125$ (parameter chosen to be tangent to experimental data at $q = 3$); the lognormal model with $\mu = 0.25$; the log-Poisson model with $c = 2$ and $\beta = 2/3$; and the log-stable model with $\alpha = 1.5$ and $C_1 = 0.15$. The three log-ID models are very close one to each other, until the moment of about 6. Larger moments are very difficult to accurately estimate and need a huge number of data points.

Here the codimension is used instead of the more frequent dimension $f(h) = D - c(h)$; for convenience we note here the singularities $h = -\gamma$ and $p(\Delta U_\ell)$ is the probability density of velocity fluctuations. The asymptotic limit of these relations is assumed to be $\ell \to 0$, hence negative singularities $h < 0$ correspond to blow up of the velocity shears. Using a saddle point argument (Parisi and Frisch, 1985), estimation of the statistical moment gives the classical Legendre transform between $\zeta(q)$ and $c(h)$:

$$\zeta(q) = \min_{h}\{qh + c(h)\} \qquad (2.73)$$

For the velocity fluctuations, most singularities are positive, and negative singularities can be obtained only if the moment function $\zeta(q)$ is decreasing, as shown by the following expression of the Legendre transform:

$$\begin{cases} h = \zeta'(q) \\ c(h) = \zeta(q) - qh \end{cases} \qquad (2.74)$$

When there is a minimum value for the singularity $h_{\min}$ (corresponding to a maximum value for the singularity of the dissipation), there is a critical moment

q_0 associated to this minimum singularity, so that for moments $q > q_0$ the structure function scaling exponent is linear:

$$\zeta(q) = c(h_{min}) + qh_{min} = \zeta(q_0) + (q - q_0)h_{min} \qquad (2.75)$$

When performing data analysis, such behavior is always found, since a minimum singularity is accessible from experimental data. For some models (e.g., log-Poisson model), this is intrinsic to the model properties, whereas for other models it will correspond to an upper-bound of the validity of the model.

In a joint study between several teams, an agreement was found until a moment of order 6, and after this experimental values were different (Arnéodo et al., 1996): this could be due to sampling limitations or could be explained by a divergence of moments of the velocity field around moment of order 6 or 7, as found in Schmitt et al. (1994).

2.6 Passive scalar intermittent turbulence

For a passive scalar, the previous approach for the velocity field may be repeated, with some important modifications. A passive scalar is a "substance" (Shraiman and Siggia, 2000), which is a scalar property of the fluid (e.g., its temperature, the salinity for marine waters, or the concentration of a chemical species) that has no effect on the dynamical transport equation. Such scalar property is hence transported by turbulence, and obeys a linear equation:

$$\frac{\partial \theta}{\partial t} + u_i \frac{\partial \theta}{\partial x_i} = \kappa \nabla^2 \theta \qquad (2.76)$$

where θ is the concentration of the passive scalar, u is the velocity field which transports the substance and obeys Navier-Stokes equations, and κ is the molecular diffusivity. In addition to the large Reynolds number, we need here to assume a large Peclet number $Pe = VL/\kappa$: this is the inertial "convective subrange," associated to large Peclet and Reynolds numbers.

In homogeneous and locally isotropic turbulence, a scaling law was put forward by Obukhov (1949) and Corrsin (1951), which writes for the passive scalar power spectrum in the inertial convective subrange:

$$E_\theta(k) \sim \epsilon^{-1/3} \chi k^{-5/3} \qquad (2.77)$$

where ϵ is the dissipation, $\chi = \kappa \langle |\nabla \theta|^2 \rangle$ is the scalar variance dissipation rate (the analogous of the dissipation, for the passive scalar). Here there is still a $-5/3$ scaling law for passive scalars; but contrary to what is found for the velocity field, there are now two dissipation quantities that are involved dimensionally: the dissipation of kinetic energy, and the dissipation of scalar variance. In real space, the equivalent relation is $\Delta\theta_\ell \sim \epsilon^{-1/6}\chi^{1/2}\ell^{1/3}$, where $\Delta\theta_\ell = |\theta(x+\ell) - \theta(x)|$ is the longitudinal increments of the Eulerian passive scalar fields at a spatial scale ℓ.

The intermittent generalization is less simple. It can be assumed that there is an energy cascade from large to small scales leading to an intermittent field ϵ_ℓ, and another cascade for the scalar variance flux from large to small scales, leading to an intermittent field χ_ℓ. The two fluxes are usually assumed to be conserved, $\langle\chi_\ell\rangle = cte$ and $\langle\epsilon_\ell\rangle = cte$. However, dimensionally the passive scalar structure functions involve a nonlinear product between the two fluxes, and without knowing the correlations between these fluxes, no simple expression can be inferred for the passive scalar exponents (Schmitt et al., 1996). Let us write the scaling expression for the passive scalar structure functions:

$$\langle\Delta\theta_\ell^q\rangle \sim \ell^{\zeta_\theta(q)} \tag{2.78}$$

with an exponent $\zeta_\theta(q)$. Due to the nonlinear product between the two fluxes, there is no fixed point like $\zeta_V(3) = 1$ for the velocity structure functions. It can only be guessed, and shown experimentally, that $\zeta_\theta(q)$ is a nonlinear and concave curve. The general form given below by Equation (2.82) can be used.

From experiments or numerical studies, it is known that the passive scalars are more intermittent than the velocity field, meaning that the deviation from the nonintermittent case ($\zeta(q) = q/3$ in both cases) is larger for the passive scalar field:

$$\zeta_\theta(q) < \zeta_V(q) \quad ; \quad q > 2 \tag{2.79}$$

Figure 2.8 shows typical estimates of the passive scalar moment function, and comparison with the velocity. Such increased intermittent behavior for the passive scalar case has been interpreted as linked to the ramp-and-cliff structures of passive scalars (Sreenivasan, 1991; Shraiman and Siggia, 2000; Warhaft, 2000; Celani et al., 2000; see Figure 2.9 for an illustration of this phenomenon). This question will be addressed in a further chapter in more detail.

2.7 Multiscaling formulation of general nonstationary scaling time series

We consider here time series. The previous theoretical approach coming from the field of turbulence was presented for a spatial scaling. The passage from spatial to temporal scale is classically done through Taylor's hypothesis. When there is a mean field $U \gg u'$, where U is the mean velocity and u' the rms (root mean square) of the fluctuations, this hypothesis states that spatial fluctuations are translated into temporal fluctuations.

In turbulence and in geosciences, many studies are performed at fixed positions, considering time statistics at a fixed location. This corresponds to an Eulerian study,

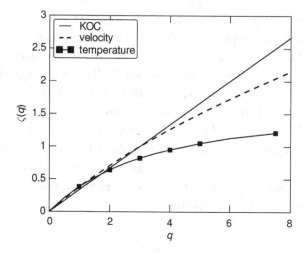

Figure 2.8 Typical $\zeta(q)$ curve obtained for the velocity and passive scalar turbulence, compared with the Kolmogorov-Obukhov-Corrsin nonintermittent curve $q/3$. For low-order moments the two curves are close and for large moments the passive scalars are much more intermittent. Adapted from Schmitt et al. (1996).

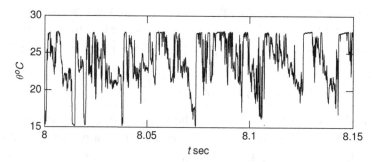

Figure 2.9 Illustration of the ramp-cliff phenomenon for the turbulent temperature.

using Taylor's hypothesis (Taylor, 1938): for a space-time process $X(x,t)$, when $U \gg u'$, the quantity $X(x, t + \tau)$ is assumed to be close to $X(x - U\tau, t)$, so that temporal and spatial increments are statistically related. In this framework, structure functions and power spectral analysis are done in the time domain, ℓ being now a time increment, and spectral analysis involves the estimation of $E(f)$, where f is the frequency.

In the general case, a nonstationary scaling time series $X(t)$ has scaling increments as follows:

$$\langle |\Delta X_\ell|^q \rangle \approx \ell^{\zeta_X(q)} \qquad (2.80)$$

for a given range of scales $\ell_0 \leq \ell \leq L$. The moment of order 1 is used to define the so-called Hurst exponent $H = \zeta_X(1)$. A multiplicative cascade process ϕ_ℓ is assumed to exist of the form

$$\phi_\ell \stackrel{d}{=} \Delta X_\ell / \ell^H \tag{2.81}$$

and the scaling exponents $\zeta_X(q)$ and $K_\phi(q)$ (where $\langle(\phi_\ell)^q\rangle \approx \ell^{-K_\phi(q)}$) are related through the relation:

$$\zeta_X(q) = qH - K_\phi(q) \tag{2.82}$$

Any cascade model, as presented in the previous section, can be used to model $K_\phi(q)$ and the multi-scale statistics of $|\Delta X_\ell|$ are then characterized by the parameter H and the different parameters used in the cascade model. This provides a characterization of the statistics of the fluctuations $|\Delta X_\ell|$ for any scale belonging to the scaling range $\ell_0 \leq \ell \leq L$ and for any moment q.

It is a multiple-scaling approach which is now very generic, and has been applied to many fields in thousands of academic studies, the root and theoretical motivation of the approach being intermittent turbulence as developed above.

2.8 Intermittency and Lagrangian turbulence

We consider here the Lagrangian case, with a referential following fluid particles. Contrary to the Eulerian cases, statistics here will depend only on time (without using Taylor's hypothesis). The Kolmogorov scale writes now $\tau_\eta = \sqrt{\nu/\epsilon}$. The scaling relations are then assumed to be valid on the range between the injection time scale T_L and τ_η.

There has been important progress in the field of Lagrangian turbulence since the beginning of the 2000s (Mordant et al., 2002; Yeung, 2002; Chevillard et al., 2003; Biferale et al., 2004; Xu et al., 2006b; Chevillard and Meneveau, 2006; Toschi and Bodenschatz, 2009; Meneveau, 2011). We consider times in the inertial range: $\tau_\eta < \tau < T_L$. In the following we note $V(t)$ and $\Theta(t)$ the Lagrangian velocity and passive scalar recorded along a fluid particle, assuming statistical homogeneity.

For Lagrangian statistics, scaling laws have been obtained dimensionnally along the same lines as for the Eulerian framework (Kolmogorov and Obukhov-Corrsin laws), for the Lagrangian velocity and passive scalar time increments $\Delta V_\tau = |V(t+\tau) - V(t)|$ and $\Delta \Theta_\tau = |\Theta(t+\tau) - \Theta(t)|$. This gives Landau's relation for the velocity (Landau and Lifshits, 1944):

$$\Delta V_\tau \sim \epsilon^{1/2} \tau^{1/2} \tag{2.83}$$

and Inoue's law for passive scalars (Inoue, 1952):

$$\Delta \Theta_\tau \sim \chi^{1/2} \tau^{1/2} \tag{2.84}$$

In the Fourier frame, there are also power law spectra, with a slope of 2:

$$E_V(f) = C_\epsilon \epsilon f^{-2} \quad ; \quad E_\Theta(f) = C_\chi \chi f^{-2} \tag{2.85}$$

where f is the frequency. Here also, velocity and passive scalar have the same scaling behavior, but with slopes of 2 instead of 5/3 for the Eulerian case.

As for the Eulerian framework, intermittency introduces corrections and new scaling expressions. Fluxes of dissipation of kinetic energy and of passive scalar variance through scale τ can be defined as $\epsilon_\tau = (\Delta V_\tau)^2/\tau$ and $\chi_\tau = (\Delta \Theta_\tau)^2/\tau$. These fluxes are assumed to be modeled by multiplicative cascades from large to small scales and produce multifractal statistics (Novikov, 1989, 1990):

$$\langle (\epsilon_\tau)^q \rangle \sim \tau^{-K_{L,\epsilon}(q)} \quad ; \quad \langle (\chi_\tau)^q \rangle \sim \tau^{-K_{L,\chi}(q)} \tag{2.86}$$

Conversely, Lagrangian structure functions for velocity and passive scalar also possess scaling statistics:

$$\langle (\Delta V_\tau)^q \rangle \sim \tau^{\zeta_{L,V}(q)} \quad ; \quad \langle (\Delta \Theta_\tau)^q \rangle \sim \tau^{\zeta_{L,\Theta}(q)} \tag{2.87}$$

and the RSH in the Lagrangian frame correspond to the following equalities between scaling exponents for velocity, passive scalar, and their fluxes:

$$\zeta_{L,V}(q) = \frac{q}{2} - K_{L,\epsilon}\left(\frac{q}{2}\right) \quad ; \quad \zeta_{L,\Theta}(q) = \frac{q}{2} - K_{L,\chi}\left(\frac{q}{2}\right) \tag{2.88}$$

We see that the theoretical expressions for Lagrangian velocity and passive scalar are very similar; there is no strong difference as in the Eulerian situation. We also see that the assumption of conservation of fluxes, corresponding to $K_{L,\epsilon}(1) = 0$ and $K_{L,\chi}(1) = 0$, lead to $\zeta_{L,V}(2) = 1$ and $\zeta_{L,\Theta}(2) = 1$ even in case of intermittency. Hence the power spectra scaling exponents are theoretically exactly 2 in each case, even when there is intermittency: in the Lagrangian framework, there is no intermittency correction for power spectral exponents.

Concerning velocity fluctuations, in the last 10 years, there has been many studies, using experiments or DNS, that have estimated the Lagrangian structure functions $\zeta_{L,V}(q)$ (Chevillard et al., 2003; Biferale et al., 2004; Xu et al., 2006a; Huang et al., 2013). For the passive scalar, there is at the present time only one study that has estimated the Lagrangian scaling exponents (Bec et al., 2014).

3

Scaling and intermittent stochastic processes

3.1 Introduction

In this section we shall consider several classical scaling stochastic processes belonging to the family of nonstationary processes with stationary increments, and recall their basic properties. They will be used in the next chapter to test and compare several methodologies to deal with scaling time series, and in the application chapters.

We shall first consider the scaling properties of the Fourier power spectrum of the time series $X(t)$, assumed to be on the form:

$$E_X(f) = Cf^{-\beta} \tag{3.1}$$

where C is a constant and $\beta > 0$ is the spectral exponent.

We shall also consider the moments of order $q > 0$ of fluctuations at scale ℓ $M_\ell(q)$ of a scaling time series $X(t)$, when they exist:

$$M_\ell(q) = \langle \Xi_\ell(X)^q \rangle \tag{3.2}$$

where ℓ is the scale belonging to a given range between a larger and a small scale, and $\Xi_\ell(X)$ is an operator extracting a characteristic fluctuation at scale ℓ of the time series $X(t)$; classically it can be the structure function $\Xi_\ell(X) = |X(t+\ell) - X(t)|$, but other methods are possible to extract stationary information from the time series, such as wavelet analysis; empirical mode decomposition; and detrended analyses (see Chapter 4). Here, for explanation purposes, we are going to use the structure functions. Since we are dealing with scaling processes that have stationary increments, we expect the following scaling behavior:

$$M_\ell(q) = A_q \ell^{\zeta(q)} \tag{3.3}$$

where A_q is a parameter independent from the scale, and $\zeta(q)$ the scaling moment function. Since the power spectrum is a second order statistics, we have the following relation between the spectral exponent β and $\zeta(2)$, when the second moment exists:

$$\beta = 1 + \zeta(2) \tag{3.4}$$

The knowledge of the full $\zeta(q)$ function then provides much more information than the single parameter β. Some completely different stochastic processes may possess the same spectral exponent, showing that, when doing data analysis and model assessment, estimation of $\zeta(q)$ on a full range of values is much better than only estimating the single parameter β. Below, when considering several scaling processes, we will provide the analytical expression of $\zeta(q)$.

3.2 Brownian motion, fractional Brownian motion, and Lévy stable walks

Here we shall consider classical linear scaling models and provide their scaling moment function.

3.2.1 Brownian motion

Let us denote $N(\mu, \sigma^2)$ the normal distribution with mean μ and variance σ^2. Then the 1D Brownian motion (or Wiener process in the mathematical literature) can be defined as a continuous stochastic process having Gaussian stationary increments such that $X(t + \ell) - X(t)$ belongs to $N(0, \ell)$.

Then, it is classically known that Equation (3.3) is verified and the scaling moment function is given by:

$$\zeta_B(q) = \frac{q}{2} \tag{3.5}$$

We also have $\beta = 2$.

3.2.2 Fractional Brownian motion

Fractional Brownian motion (fBm), a generalization of Brownian motion, was introduced in 1940 by Kolmogorov (Kolmogorov, 1940); it was more extensively studied by Mandelbrot and coworkers in the 1960s (Mandelbrot and Van Ness, 1968) and since then, it has been considered as a classical scaling stochastic process.

A possible definition is the following: for $0 \leq H \leq 1$, a fractional Brownian motion $B_H(t)$ of parameter H is defined as a Gaussian process of mean 0 and covariance $\langle B_H(t) B_H(s) \rangle = \frac{1}{2}(|t|^{2H} + |s|^{2H} - |t - s|^{2H})$. Then it is easily shown that Equation (3.3) is verified and the scaling moment function is given by:

$$\zeta_H(q) = qH \tag{3.6}$$

3.2 Brownian motion, fractional Brownian motion, and Lévy stable walks

For $H = 1/2$ the Brownian motion is recovered and when $0 \leq H < 1/2$, increments are negatively correlated, whereas when $1/2 \leq H < 1$ increments are positively correlated. The fBm (H) can also be constructed explicitly using a stochastic integral representation (Embrechts and Maejima, 2002):

$$B_H(t) = \int_{-\infty}^{0} \left((t-u)^{H-1/2} - (-u)^{H-1/2}\right) dB(u) + \int_{0}^{t} (t-u)^{H-1/2} dB(u) \quad (3.7)$$

We also have $\beta = 1 + 2H$, so that time series modeled by fBm have a spectral exponent between 1 and 3.

3.2.3 Lévy stable motion

Let us recall that Lévy stable laws are generalizations of the Gaussian law. Named after the French mathematician Paul Lévy (1886–1971), these laws are stable and attractive under addition (basic references are Gnedenko and Kolmogorov, 1954; Zolotarev, 1986; Samorodnitsky and Taqqu, 1994; Janicki and Weron, 1994; Nikias and Shao, 1995).

One possible definition is the following. A random variable X is stable if, for any integer n, and for any independent copy X_1, X_2, .., X_n of X, there exist two reals, $C_n > 0$ and D_n such that there is the equality in distribution:

$$X_1 + X_2 + ... + X_n \stackrel{d}{=} C_n X + D_n, \quad (3.8)$$

When $D_n = 0$ the law is called strictly stable. These laws depend on a basic parameter, the Lévy stable index α: $C_n = n^{1/\alpha}$. This parameter is bounded between 0 and 2. For $\alpha = 2$ one recovers the Gaussian law. For $\alpha < 2$, there is no general closed analytical form to express the probability density of such random variables. However, the second Fourier characteristic function $\Psi_X(u)$, is known:

$$\Psi_X(u) = \log\langle e^{iuX} \rangle = -\sigma^\alpha |u|^\alpha \left(1 - i\beta(\text{sign}(u))\tan\frac{\pi\alpha}{2}\right) + i\mu u. \quad (3.9)$$

The parameters are the index of stability, $0 < \alpha \leq 2$, the scale parameter $\sigma \geq 0$, the skewness parameter $-1 \leq \beta \leq 1$, and the shift parameter $\mu \in \mathcal{R}$.

Similarly to the Gaussian noise, Lévy stable noise can be defined and, as the Brownian motion is the integral of a Gaussian noise, Lévy motion is the integral of a Lévy noise. Lévy motion should not be confused with Lévy flight, which is usually defined as a random walk with step length having a heavy tailed distribution.

For $\alpha < 2$ Lévy distributions are heavy tailed, meaning that extremes are much more frequent than for a Gaussian distribution: the moments of order α diverge: for $q < \alpha$, $\langle |X|^q \rangle < \infty$ and $\langle |X|^\alpha \rangle = \infty$. This is one of the main properties of Lévy stable random variables and also the main difference with the Gaussian distribution and all other distributions with exponential or stretched exponential decrease.

Lévy stable motions are also scaling, but because of the divergence of moments of order α, the moment function is peculiar. For $q \le \alpha$, one has $H = 1/\alpha$ and the moment function is linear of the form $\zeta_L(q) = \frac{q}{\alpha}$. However, for $q > \alpha$, the consequence of the divergence of moments is such that, the estimated function depends on the number of realizations. As shown in Schmitt et al. (1999) and also by Nakao (2000), for one realization the moment function is bilinear:

$$\zeta_L(q) = \begin{cases} \frac{q}{\alpha}; & q \le \alpha \\ 1; & q > \alpha \end{cases} \quad (3.10)$$

this saturation for moments $q > \alpha$ comes from a cancellation of the effect of integration and of divergence of moments.

The power spectrum is not considered for such processes, since this corresponds to a second order moment, and when $\alpha < 2$, the second moment is not defined.

3.2.4 Fractional Lévy stable motion

As shown above for the fractional Brownian motion, a fractional Lévy motion $L_H(t)$ of order $0 \le H \le 1$ can be introduced to generalize Lévy stable motion (Taqqu and Wolpert, 1983; Maejima, 1983; Taqqu, 1988; Embrechts and Maejima, 2002; Samorodnitsky and Taqqu, 1994). One of the possible definitions involves a stochastic integral (Samorodnitsky and Taqqu, 1994):

$$L_{H,\alpha}(t) = \int_{-\infty}^{\infty} \left(|t - u|^{H-1/\alpha} - |u|^{H-1/\alpha} \right) dL(u) \quad (3.11)$$

As for the Lévy stable motion, moments of order $q < \alpha$ have a linear moment function and larger moments have a different expression, due to the divergence of moments. The corresponding bilinear expression is now written for 1 realization (Schmitt et al., 1999):

$$\zeta_{L,H}(q) = \begin{cases} qH; & q \le \alpha \\ q\left(H - \frac{1}{\alpha}\right) + 1; & q > \alpha \end{cases} \quad (3.12)$$

As for the Lévy stable motion, the change of slope between the two portions of straight lines is $\frac{1}{\alpha}$.

3.2.5 Self-similar processes

Such processes (Brownian motion, fractional Brownian motion, Lévy motion, and fractional Lévy motion) are often called "self-similar processes" (Embrechts and Maejima, 2002; Samorodnitsky and Taqqu, 1994). A process $X = \{X(t), t \in R\}$ is self-similar with index H, if for any $\lambda > 0$:

$$\{X(\lambda t)\} \stackrel{d}{=} \lambda^H \{X(t)\} \quad (3.13)$$

3.3 Simulation of continuous multifractal processes

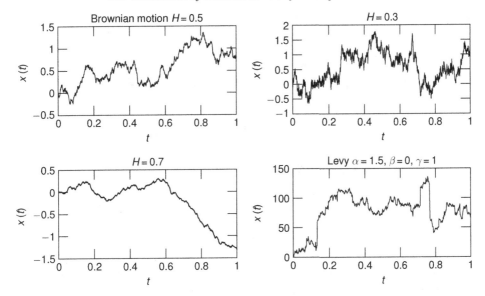

Figure 3.1 Simulations of various scaling processes (Brownian motion, fractional Brownian motion, Lévy stable motion).

These self-similar processes have stationary increments; those of Brownian motion and Lévy motion have independent increments; for fractional Brownian motion and fractional Lévy motion, the increments display long-range dependence and are called "long-memory processes" (Beran, 1994). We have reviewed here only some self-similar processes; there are many other self-similar processes, such as log-fractional stable motion, harmonizable stable motion, subordinated processes, etc. For a much more complete description, see Embrechts and Maejima (2002) and Samorodnitsky and Taqqu (1994).

Figure 3.1 shows several simulations of Brownian motion, fractional Brownian motion and Lévy stable motion. Figure 3.2 shows typical moment functions $\zeta(q)$ for several cases, Brownian motion, fractional Brownian motion for $H = 0.3$, Lévy stable motion with $\alpha = 1.5$, and a multifractal time series with a nonlinear moment function (see next subsection).

3.3 Simulation of continuous multifractal processes

All the models presented in the previous sections (from Brownian motion to fractional Brownian motion, and Lévy and fractional Lévy motions) are linear, belonging to self-similar processes. The distributions moment functions obey the simple rescaling form of Equation (3.13), which depends only on one parameter $0 \leq H \leq 1$. The corresponding moments functions are linear or bilinear. They may be called monofractal or bi-fractal processes. Let us note, however, that in the

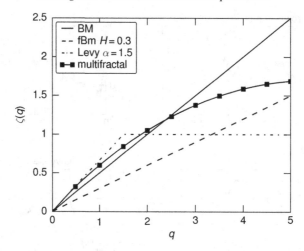

Figure 3.2 Typical moment functions $\zeta(q)$ for several cases, Brownian motion, fractional Brownian motion for $H = 0.3$, Lévy stable motion with $\alpha = 1.5$ and a multifractal time series with a nonlinear moment function.

mathematical literature, the Lévy motions are sometimes considered to be multifractal since a bilinear expression is not linear (Jaffard, 1999). However, the underlying model is still linear, and here we shall not be considering these models in the category of multifractal models as they rely on a completely different construction.

Here we shall consider models to generate multifractal processes of the form given by Equation (3.3), with a nonlinear moment function $\zeta(q)$, which we can write using the following general form (Equation [2.82]):

$$\zeta(q) = qH - \Psi(q) \qquad (3.14)$$

where the parameter $0 \leq H \leq 1$ is given by

$$H = \zeta(1) \qquad (3.15)$$

and $\Psi(q)$ contains the nonlinear term. We have $\Psi(0) = 0$, $\Psi(1) = 0$ and $\Psi(q)$ is a convex function, whereas $\zeta(q)$ is concave. This was first put forward for turbulence in 1962 by Kolmogorov and Obukhov, and has been shown experimentally since 1984 (Anselmet et al., 1984). However, the construction of a stochastic process able to reproduce such property has only been done since the 1990s.

This property has been given several denominations in the last thirty years. It has been called "nonstationary" multifractals (Schertzer et al., 1997), "nonconservative" multifractals (Schertzer and Lovejoy, 1987), "multiaffine" field (Vicsek and Barabasi, 1991; Barabási et al., 1991; Kuramoto et al., 1998), and more recently, "multifractal random walk" (Bacry et al., 2001; Schmitt and Seuront, 2001). We shall use here the name "multifractal random walk," since it has gained popularity in the fields of continuous simulations and financial modelling using multifractal models.

The first proposals, in the 1990s, were discrete constructions (Barabási et al., 1991; Vicsek and Barabasi, 1991; Eggers and Grossmann, 1992; Benzi et al., 1993b; Juneja et al., 1994; Biferale et al., 1998). They mostly consisted of taking the modeled nonstationary field (e.g., turbulent velocity) as a sum of a multiplicative and correlated positive field, and a random signed term. These models did not have straightforward continuous expressions. In 2001 a model was proposed, mainly with an objective of financial applications, called "multifractal random walk" (Bacry et al., 2001). It was mainly related to the proposal of Eggers and Grossmann (1992): the discrete sum of a product of Gaussian terms (to express the sign of the fluctuations) and of lognormal terms, with the correct correlation properties in order to be the result of a multiplicative cascade. The multifractal random walk was the continuous limit of this discrete sum. The objective was to model financial fluctuations, but this process had more general applications, and in all fields with all values of H.

The first fully continuous (log-ID, see Chapter 2) multifractal constructions were published in the late 1990s and early 2000s: there has been several proposals, belonging to two different families, to generate positive multifractal measures or multifractal random walks using continuous stochastic processes. We will discuss these proposals in the following three subsections.

3.3.1 Continuous multiplicative cascades

The first step is to generate a stationary positive random measure as a continuous log-ID cascade. We do this by firstly introducing an infinitely divisible (ID) additive independently scattered random measure, called below "ID random measure" (Rajput and Rosinski, 1989; Samorodnitsky and Taqqu, 1994; Applebaum, 2004).

A (nonrandom) control measure m associates a set A to a positive number $m(A)$ with the property of additivity for disjoint sets: for sets A and B with $A \cap B = \emptyset$, we have $m(A \cup B) = m(A) + m(B)$. An ID random measure M with a so-called control measure m is the following: it associates to a set A, a random variable $M(A)$. Furthermore, $M(A)$ belongs to an ID distribution which is characterized by $\Psi(q)$ and $m(A)$. More precisely, the ID random variable $M(A)$ is characterized by its second Laplace characteristic function $\Psi_{M(A)} = m(A)\Psi(q)$ (which is assumed here to converge):

$$\langle e^{qM(A)} \rangle = e^{\Psi(q)m(A)}. \tag{3.16}$$

The analytic shape of $\Psi(q)$ characterizes the ID distribution underlying the ID random measure. We choose to normalize this function in order to have $\Psi(1) = 0$.

An important property of an *ID* random measure is the additivity property for disjoint sets: for sets A and B with $A \cap B = \emptyset$, we have $M(A \cup B) = M(A) + M(B)$.

The first use of ID random measures to model continuous multiplicative cascades was proposed by Schmitt and Marsan (2001): continuous multifractal cascades are modeled as the exponential of an ID random measure on a cone. It shows that this produces multifractal statistics, and that the long-range correlation properties are characterized by the intersection of cones (see Figure 3.3). A cone was also later and independently proposed, in the framework of compound Poisson processes, by Barral and Mandelbrot (2002). They suggested the following measure on the time-scale half-plane $\{(t, \ell); \ell > 0\}$:

$$m(dt, d\ell) = \frac{dtd\ell}{\ell^2} \tag{3.17}$$

Later, Bacry and Muzy (Muzy and Bacry, 2002; Bacry and Muzy, 2003) mixed the two previous works and published a comprehensive mathematical model for continuous log-ID cascades as the exponential of an ID measure on a cone. To recognize all contributions, and to give due recognition to the original proposal, this model is called here SMBM2 (Schmitt-Marsan-Barral-Mandelbrot-Bacry-Muzy).

The SMBM2 model is the following: The positive ID multifractal field $\epsilon(t)$ at time t is defined as the exponential of the ID random measure on a set $A_\ell(t)$:

$$\epsilon_\ell(t) = Ce^{M(A_\ell(t))} \tag{3.18}$$

where $\ell \ll 1$ is the resolution, C is a constant and $A_\ell(t)$ is a cone centered around t (Figure 3.3):

$$A_\ell(t) = \{(x, y) : \ell \leq y \leq L; \ t - y \leq x \leq t\} \tag{3.19}$$

and the associated control measure is given by Equation (3.17). Figure 3.3 shows two domains: on the left, a causal domain (times $x > t$ are not influencing the process $\epsilon_\ell(t)$); and on the right, a symmetric domain that produces a noncausal process. Equation (3.19) corresponds to the causal domain. Here, the integration domain is 2D for generating a time series; this approach can be generalized for D-dimensionnal processes, with a integration cone belonging to a dimension $D+1$ (Chainais, 2006; Schmitt and Chainais, 2007).

Here, we consider small values of ℓ and large scale ratios $\lambda = L/\ell$. By simply volume integrating of the cone, we have:

$$m(A_\ell(t)) = \int_\ell^L \frac{dy}{y^2} \int_{t-y}^t dx = \int_\ell^L \frac{dy}{y} = \log \frac{L}{\ell} \tag{3.20}$$

3.3 Simulation of continuous multifractal processes

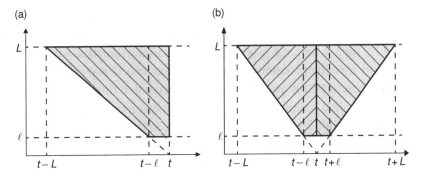

Figure 3.3 Cone $A_\ell(t)$ chosen for the stochastic integration in the $\{(t, \ell), \ell > 0\}$ plane: (a) the causal case, and (b) the symmetric noncausal case.

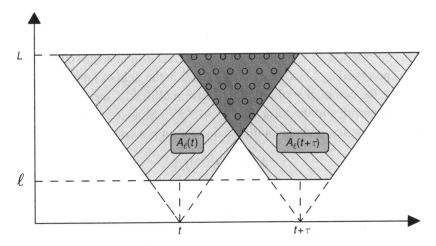

Figure 3.4 Intersection between two cones A_ℓ and $A_\ell(t + \tau)$. The darker zone corresponds to the correlation between the two points $X(t)$ and $X(t + \tau)$; it decreases as a power law and generates the long-range correlation of such process.

Hence the constructed field $\epsilon_\ell(t)$ possesses scaling statistics (Schmitt and Marsan, 2001; Muzy and Bacry, 2002):

$$\langle (\epsilon_\ell(t))^q \rangle \sim \langle \exp(qM(A_\ell(t))) \rangle \sim e^{\Psi(q)m(A_\ell)} \sim \left(\frac{L}{\ell}\right)^{\Psi(q)} \tag{3.21}$$

By construction, the stochastic process $\epsilon_\ell(t)$ also possesses long-range power-law correlations coming from the conic structure of the set A_ℓ (Schmitt and Marsan, 2001). Let us consider a time t and $\tau > 0$. The cones $A_\ell(t)$ and $A_\ell(t + \tau)$ have an intersection, which decreases with the distance τ between the two points (Figure 3.4). The long-range correlation of the process is related to the area of the intersection.

The conic structure of the integration domain contains the continuity of the cascade: every scale is involved, between the outer scale L and the small scale ℓ. Consequently, such a cascade produces scale-invariant statistics with no specific scale ratio, and are called continuous (in scale) cascades. This produces continuous multiplicative cascades, for any ID model. The generated time series is stationary, belonging to so-called multifractal measures. Such continuous cascade will be used below to generate multifractal random walks, also called nonstationary multifractal fields, or multiaffine time series. Let us recall that the ID model which is chosen is contained in the control measure chosen (Equation [3.16]): for a lognormal model the control measure is Gaussian; for a log-Poisson model the control measure is Poissonian.

3.3.2 Subordination

The first construction of a nonstationary multifractal field with stationary increments was proposed by Mandelbrot et al. (1997) in the field of finance. Let $\Theta(t) = \int_0^t \epsilon(u) du$ be the cumulative of a multiplicative multifractal measure, where ϵ has multiscaling statistics in the form of $\langle \epsilon_\ell^q \rangle \sim \ell^{-K(q)}$. Since ϵ is positive, $\Theta(t)$ is an increasing time series, and the multifractal proposal is a compound process:

$$X(t) = B_h(\Theta(t)) \qquad (3.22)$$

where $B_h(t)$ is a fractional Brownian motion of parameter h ($0 \leq h \leq 1$) and X is built using a subordination. It can be shown that Equation (3.3) is verified and the scaling exponent is (Mandelbrot et al., 1997):

$$\zeta_X(q) = qh - K(qh) \qquad (3.23)$$

Since $K(1) = 0$, one has here:

$$\zeta_X(1/h) = 1 \qquad (3.24)$$

For example, for $h = 1/2$, this gives $\zeta_X(2) = 1$. This model was first proposed in the field of finance, but it can also be used for turbulence (with $h = 1/3$), or other multiscaling time series. We can see that:

$$H = \zeta_X(1) = h - K(h) > h \qquad (3.25)$$

This shows that the classical Hurst index, H, defined as $H = \zeta_X(1)$, is different from h, hence the different notation used here. Examples of subordinated simulations of a multifractal process, following Equation (3.22), with a lognormal multiplicative cascade with $\mu = 0.2$, and values of $h = 0.3$ and $h = 0.5$ are shown in Figure 3.5.

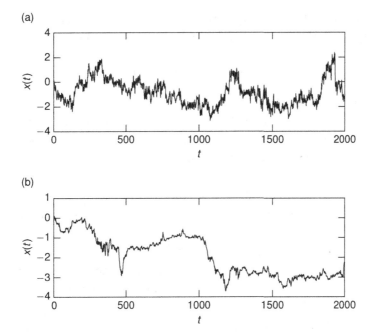

Figure 3.5 Examples of subordinated simulations of a multifractal process, following Equation (3.22), with a lognormal multiplicative cascade with $\mu = 0.2$, and values of $h = 0.3$ (above) and $h = 0.5$ (below).

3.3.3 Moving average with Brownian or fractional Brownian noises

Moving average with a multiplicative cascade

Another approach to generate a continuous multiaffine or multifractal random walk is to consider moving averages with a scaling kernel, the latter being generated using a continuous cascade framework. The first proposal was formulated by Schmitt and Seuront (2001), and Muzy and Bacry first suggested a moving average with respect to the Wiener process (Muzy and Bacry, 2002; Bacry and Muzy, 2003):

$$X_\ell(t) = \int_0^t e^{M(A_\ell(u))} dB(u) \qquad (3.26)$$

where $A_\ell(t)$ is the same as discussed in the previous subsection, for a given finite scale ℓ. Then the continuous process $X(t)$ is taken as the limit $X(t) = \lim_{\ell \to 0} X_\ell(t)$. In fact, this process can be shown to be the same as the subordinated process given by Equation (3.22) (Muzy and Bacry, 2002).

This can be generalized by taking an integrand which is a fBm with a parameter $h \neq 1/2$. In this case, the subordination construction and the moving average are producing different stochastic processes (Muzy and Bacry, 2002; Abry et al., 2009). Let us imagine, as in Abry et al. (2009) a process of the form:

$$X(t) = \int_0^t \epsilon(u) dY_h(u) \quad (3.27)$$

where $\epsilon(u)$ is a multifractal noise and $Y_h(t)$ is a self-similar process of parameter h, independent of ϵ. Abry et al. (2009) performed some simulations and showed that when $h \neq 0$ and $h > 1/2$, the subordination and the moving average methods do not generate the same type of multifractal time series: they have different multifractal properties, different generating structure, and different apparent sample paths. In the subordination case, high values of ϵ, corresponding to local high intermittency, lead to a faster time. In the moving average case, high values of ϵ translate into local large variability in the time series.

In the last decade, only a few papers have considered the mathematical construction, existence, and convergence of such a process. When taking Y_h in Equation (3.27) as a fractional Brownian motion, the first point is to construct a stochastic integral with respect to fBm, and show that it is well defined and not diverging. This was done in Ludena (2008) for $h > 1/2$ using the construction of the stochastic integral with respect to fBm presented in Pipiras and Taqqu (2000). Abry et al. (2009) further explored the case $h > 1/2$ using fractional Wiener integrals. In such a situation, the process generated is shown to be converging and different from that previously produced. Let us note $K(q)$ the scaling moment function of ϵ and $\mu = K(2)$. The result of Abry et al. (2009), with our notations, is the following (for $h > 1/2$ only):

$$\zeta_X(q) = \begin{cases} qh - K(q); & \mu \leq 2h - 1 \\ \frac{\mu+1}{2}q - K(q); & 2h < \mu + 1 \end{cases} \quad (3.28)$$

where $K(q)$ is the scaling exponent for the kernel function. When $\mu \leq 2h - 1$, we have $H = \zeta_X(1) = h$, whereas for $2h < \mu + 1$, $H = \zeta_X(1) = \frac{\mu+1}{2}$ and is not related to h.

The case $h < 1/2$ has only been considered, up to now, in one paper. In the case $h < 1/2$, Perpete (2013) has shown that the process is well defined and, using a different method, found the following scaling exponents:

$$\zeta_X(q) = \frac{q}{2} - K(q) \quad (3.29)$$

with the following conditions:

$$\begin{cases} 0 < h < 1/4 \ \& \ K(\frac{1}{2h}) < \frac{1}{2h} - 1 \\ 1/4 < h < 1/2 \end{cases} \quad (3.30)$$

This result is surprising since there is no h dependence in the value of $\zeta_X(q)$. We have here in both cases $H = \zeta_X(1) = 1/2$. The different H values are shown in Figure 3.6 in a $h - \mu$ plane.

3.3 Simulation of continuous multifractal processes

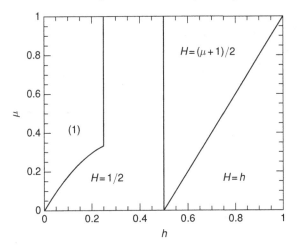

Figure 3.6 Different values of $H = \zeta_X(1)$ obtained in a $h - \mu$ plane. The region (1) corresponds to a region with unknown value for H. The nonlinear curve which is at the bottom left corresponds to the condition $K(1/2h) < 1/2h - 1$.

Examples of moving average simulations of a multifractal process, following Equation (3.27) with a lognormal multiplicative cascade with $\mu = 0.2$, and values of $h = 0.3$ and $h = 0.5$, are shown in Figure 3.7.

In the lognormal case, since $K(q) = \frac{\mu}{2}(q^2 - q)$, the condition $K(\frac{1}{2h}) < \frac{1}{2h} - 1$ becomes $\mu < 4h$. Since $0 < h < 1$ and $0 < \mu < 1$, we can plot in the $h - \mu$ plane the result in Figure 3.8, following the calculations of Abry et al. (2009) and Perpete (2013), given in Equations (3.28) and (3.30). There are four zones in this $h - \mu$ quadrant: from left to right, there is no result in the zone left blank; for the next region we have $\zeta_X(q) = \frac{q}{2} - K(q)$; then $\zeta_X(q) = \frac{\mu+1}{2}q - K(q)$ and finally $\zeta_X(q) = qh - K(q)$.

Generalization: moving average with the power $a > 0$ of a multiplicative cascade

There is a way to change the value of $\zeta_X(q)$, by taking a power $a > 0$ of the multiplicative cascade in the stochastic integral:

$$X(t) = \int_0^t \epsilon^a(u) dY(u) \tag{3.31}$$

The exponent $\zeta_X(q)$ is modified as follows (Abry et al., 2009; Perpete, 2013):

$$\zeta_X(q) = \begin{cases} \frac{q}{2} - K(aq); & 0 < h < 1/4 \ \& \ K(\frac{1}{2h}) < \frac{1}{2h} - 1 \\ \frac{q}{2} - K(aq); & 1/4 < h < 1/2 \\ qh - K_2(q,a); & 1/2 < h \ \& \ K_2(2,a) \leq 2h - 1 \\ \frac{\mu+1}{2}q - K_2(q,a); & 1/2 < h \ \& \ 2h < K_2(2,a) + 1 \end{cases} \tag{3.32}$$

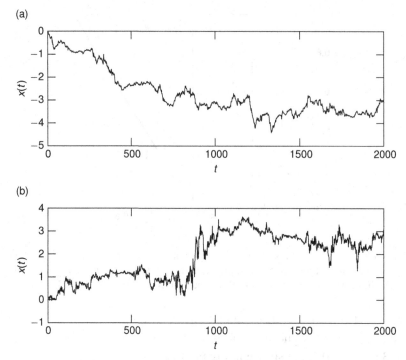

Figure 3.7 Examples of moving average simulations of a multifractal process, following Equation (3.27) with a lognormal multiplicative cascade with $\mu = 0.2$, and values of $h = 0.3$ (above), and $h = 0.5$ (below).

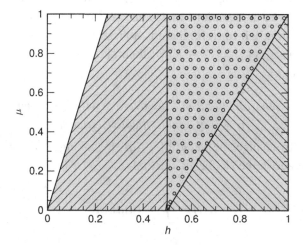

Figure 3.8 The plane $h - \mu$ giving the value of the scaling exponent $\zeta_X(q)$ for a lognormal process, according to Abry et al. (2009) and Perpete (2013). For the zone which is left in blank, there is no result for the moment, the convergence is not yet demonstrated. Increasing diagonal: $\zeta_X(q) = \frac{q}{2} - K(q)$; open dots: $\zeta_X(q) = \frac{\mu+1}{2}q - K(q)$ and decreasing diagonals $\zeta_X(q) = qh - K(q)$.

where $K_2(q, a) = K(aq) - qK(a)$. For a lognormal process, this gives $K_2(q, a) = a^2 K(q)$.

Let us consider the following quadratic form of the scaling moment function $\zeta(q)$:

$$\zeta_X(q) = qH - \frac{\mu'}{2}(q^2 - q) \qquad (3.33)$$

where the parameters are $H = \zeta_X(1)$ and $\mu' = 2H - \zeta_X(2)$. This is a general expression of a quadratic moment function, corresponding to a lognormal process characterized by the parameters H and μ'. Our question here is: Is it possible to retrieve Equation (3.33) through a stochastic moving average of the form Equation (3.27), and if yes, for which values of h, a, and μ (expressed using the desired parameters H and μ')?

- Case $H < 1/2$: We need to take $h > 1/2$; in this case the function write $\zeta_X(q) = \frac{q}{2} - K(aq)$ and does not depend on h. This gives the values:

$$a = \frac{\mu'}{2H - 1 + \mu'}; \quad \mu = \frac{\mu'}{a^2} \qquad (3.34)$$

where we need to have $\mu' > 1 - 2H$. Thus all values are not accessible: the smaller H, the more the process must be intermittent (large μ').

- Case $H > 1/2$: Considering $a < 1$, the previous case with $h < 1/2$ will provide $H = 1/2 - K(a) > 1/2$. Hence Equation (3.34) is still valid.

  Or by taking $h > 1/2$, if $\mu' < 2H - 1$ we have $h = H$. In this case any a and μ are possible with the condition $a^2 \mu = \mu'$.

  Finally if $\mu' \geq 2H - 1$, we have:

$$\mu = 2H - 1; \quad a = \sqrt{\frac{\mu'}{2H - 1}} \qquad (3.35)$$

Thus in the case $H > 1/2$, the simulation is possible for all couple of values of (H, μ').

As a conclusion, we may consider that multifractal random walks (or nonstationary multifractals, or multi-affine fields) have been obtained for a long time in many fields of applied and natural sciences. However, to generate such time series has long been a challenge. We have reviewed here some recently obtained results and considered the situation of a desired quadratic moment function (for the lognormal case taken as first example) characterized by the parameters H and μ', and shown how to choose the parameters h, a, and μ to retrieve this quadratic curve. Depending on the values of H and μ', some values are still not accessible, showing that a more general stochastic generating equation is still needed: Equation (3.31) is not the final answer.

4

New methodologies to deal with nonlinear and scaling time series

As mentioned previously, turbulent flows or turbulent-like phenomena possess in many aspects self-similarity or multiscaling properties and multiscale statistics. The classical way to characterize this scaling behavior is to estimate the qth-order structure function $S_q(\ell)$ for different scales. Power-law behavior is then expected, e.g., $S_q(\ell) \sim \ell^{\zeta(q)}$, in which $S_q(\ell)$ represents the qth-order structure function (see Chapter 2). The structure function method was first proposed by Kolmogorov in 1941 in his famous homogeneous and isotropic turbulence paper. There are also to be found other methodologies to characterize the multiscale statistics of such fields. Such examples include: wavelet-based approaches; the Hilbert-Huang transform; detrended fluctuation analysis; and detrended structure function analysis, to mention a few. Each method has its own advantages and disadvantages as it might show different performances for different types of data. Before applying one of these methods to real data, it is necessary to know whether the method is suitable for that data. For example, in the fluid geosciences, the annual cycle is present in many observation data, and therefore the structure function analysis, as we show below, might not be suitable. Here, other methodologies which might constrain the large-scale influence should be considered.

In this chapter, we will present more details about the classical structure-function analysis and new approaches that have emerged in recent years. This chapter is organized as follows. The increment-based methods are introduced in Sections 4.1 to 4.3. The wavelet-based methods are presented in Section 4.7. The Hilbert-based method is described in Section 4.8.

4.1 Structure function analysis

4.1.1 Second-order structure function

We first consider here the second-order structure functions (SFs). The second-order SF is defined as,

4.1 Structure function analysis

$$S_2(\ell) = \langle \Delta_\ell u(x)^2 \rangle, \quad \Delta_\ell u(x) = u(x+\ell) - u(x) \tag{4.1}$$

in which ℓ is a separation scale (Figure 4.1). Originally, such an approach was done for spatial scales; however, as explained in Chapter 2, we are dealing here with time series and transform spatial fluctuation into time fluctuation using Taylor's frozen hypothesis, hence, here ℓ is a time scale. For a scaling process, a power-law behavior is expected,

$$S_2(\ell) \sim \ell^{\zeta(2)} \tag{4.2}$$

in which $\zeta(2)$ is the scaling exponent. It could be further associated to the Fourier power spectrum of $u(x)$ by using the Wiener-Khinchin theorem,

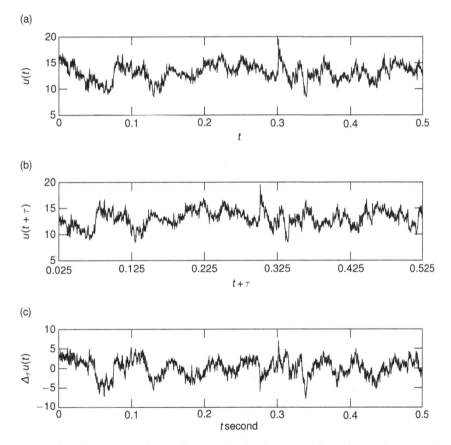

Figure 4.1 Illustration of the turbulent velocity increment $\Delta_\tau u(t) = u(t) - u(t+\tau)$ obtained from a wind tunnel experiment with $Re_\lambda \simeq 720$ and a time increment $\tau = 0.025$ sec; a) A portion of 0.5 second measured longitudinal velocity $u(t)$; b) A time $\tau = 0.025$ second shift $u(t+\tau)$; c) The measured velocity increment $\Delta_\tau u(t)$. In the context of turbulence, the separation time scale τ is interpreted in the inertial range as the typical time scale of a vortex with time scale τ.

$$S_2(\ell) = \int_0^{+\infty} E_\Delta(f)df = \int_0^{+\infty} E(f)\left(1 - \cos(2\pi f\ell)\right)df \tag{4.3}$$

in which $E(f)$ is the Fourier power spectrum of $u(t)$ and is assumed to have a power law form,

$$E(f) = cf^{-\beta} \tag{4.4}$$

where c is a constant. Note that in Equation (4.3) the integral constant has been ignored. A convergence condition requires that β should be in the range $1 < \beta < 3$. A scaling transform leads to an analytical solution (Huang et al., 2010),

$$S_2(\ell) = \frac{c\pi^{\beta-\frac{1}{2}}\Gamma(\frac{3}{2}-\frac{\beta}{2})}{(\beta-1)\Gamma(\frac{\beta}{2})}\ell^{\beta-1} \tag{4.5}$$

in which Γ is the Gamma function, resulting in a scaling relation,

$$\zeta(2) = \beta - 1 \tag{4.6}$$

Note that Equation (4.3) indicates that all Fourier components, except for $f = n/\ell$, $n = 0, 1, 2 \cdots$ have contribution to $S_2(\ell)$. We also note that the Fourier power spectrum is decaying with f, implying that the contribution from the low frequency part (large scale part) might be important.

4.1.2 Weight-function and cumulative function

To understand more quantitatively the contribution from different Fourier modes, we introduce here a weight-function $\mathcal{W}(\ell,f)$ defined by

$$S_2(\ell) = \int_0^{+\infty} E_u(f)\mathcal{W}(\ell,f)df \tag{4.7}$$

The weight function characterizes the contribution of a Fourier component in the estimation of the second-order structure-function (Huang et al., 2010). Figure 4.2 shows $\mathcal{W}(\ell,f)$ for the second-order structure-function for two separation scales ℓ and 2ℓ. The overlap of $\mathcal{W}(\ell,f)$ and $\mathcal{W}(2\ell,f)$ indicates that the Fourier components have contribution for both these two scales. This means that the weight function to estimate $S_2(\ell)$ and $S_2(2\ell)$ are not localized in the frequency space: they have important overlaps, showing that the spatial structure function is not associated to fluctuation at a given frequency, hence, a given time scale (as often implicitly assumed). This is a "mixture" of scales in the estimation of structure function, hence an artificial correlation between two different scales (Huang et al., 2014).

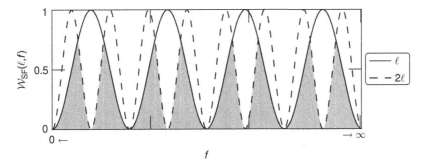

Figure 4.2 Illustration of weight-function $\mathcal{W}(\ell,k)$ for two different separation scales ℓ and 2ℓ. Note that except for the case $f_\ell = n/\ell$ with $n = 0,1,2,\cdots$ all Fourier components have contributions to $S_2(\ell)$.

Let us introduce a relative cumulative function as follows:

$$\mathcal{P}(\beta,f) = \frac{\int_0^f E_u(k')(1-\cos(2\pi k'))dk'}{\int_0^{+\infty} E_u(k')(1-\cos(2\pi k'))dk'} \times 100\% \quad (4.8)$$

It provides a measurement of the relative contribution from the frequency band $[0,f]$. An analytical expression of Equation (4.8) for a pure power-law, $E_u(k) \sim k^{-\beta}$ is written as:

$$\mathcal{P}(\beta,f) = \frac{1}{a(\beta)}\left\{(3-\beta)(\cos(f)-1)f^{1-\beta} + g(f,\beta)f^{3-\beta}\right\} \times 100\% \quad (4.9)$$

in which $a(\beta) = \sqrt{\pi}\,(3-\beta)\,2^{1-\beta}\,\Gamma(3/2-\beta/2)\Gamma(\beta/2)^{-1}$, and $g(f,\beta) = {}_1F_2(3/2-\beta/2, 3/2, 5/2-\beta/2, -f^2/4)$ is a generalized hypergeometric function (Huang et al., 2010). Figure 4.3 a shows the measured $\mathcal{P}(\beta,k)$ for several values of β. They are $\beta = 1.4$ for active scalar in Rayleigh-Bénard convection system, $\beta = 5/3$ for 3D homogeneous and isotropic turbulence, $\beta = 2$ for Lagrangian turbulence in 3D homogeneous and isotropic turbulence, and $\beta = 3$ for the forward enstrophy cascade in 2D turbulence. Since we have the rescaling relation in Equation (4.5), we shall consider here only the case $\ell = 1$, and in particular we are concerned by the case $\mathcal{P}_1(\beta) = \mathcal{P}(\beta,f)|_{f=1}$. This represents the influence of large-scales, i.e., $f < 1$. Figure 4.3 b) displays the measured $\mathcal{P}_1(\beta)$. It increases with β as shown above, implying that the larger the parameter β, the more important influence large scales have in the structure function estimation. Therefore, this question should receive more attention when it is applied to the Lagrangian turbulence with $\beta = 2$, or the two-dimensional turbulence with $\beta = 3$, etc.

This indicates also that the structure function method might fail to detect the right scaling behavior when an energetic structure is present. This is the case for the large-scale circulation in the Rayleigh-Bénard convection system. It is also

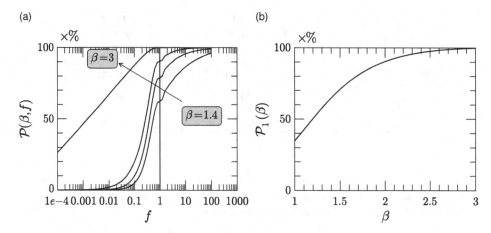

Figure 4.3 a) Measured cumulative function $\mathcal{P}(\beta,k)$ for $\beta = 1.4, 5/3, 2$, and 3. The separation scale considered here is $\ell = 1$. b) The measured large-scale influence $\mathcal{P}_1(\beta) = \mathcal{P}(\beta,1)$. Note that $\mathcal{P}_1(\beta)$ increases with β.

the case for many data sets collected from fluid geosciences with a strong annual cycle (Huang et al., 2009a), for the passive scalar turbulence with the ramp-cliff structures (Huang et al., 2010), and for vortex trapping events in the Lagrangian turbulence (Huang et al., 2013), to name a few.

4.1.3 Influence of a single scale

As we mentioned in the previous section, the structure function analysis might be influenced by energetic structures, such as the annual cycle in many fluid geosciences data. We perform here a stochastic simulation to mimic this effect. We perturbate the original spectrum $E_u(k)$ with 10% total energy, at $k = 0.001$, corresponding to $\ell = 1000$. The second-order SFs are then estimated by using the Equation (4.3) for several values of β. Figure 4.4a shows the measured second-order SFs $S_2(\ell)$ with and without perturbation for the same β as shown in Figure 4.3. To quantify the relative perturbation, a disturbance function is defined as:

$$\mathcal{I}(\beta,\ell) = 20 \times \log_{10}\left(\frac{S_2^P(\beta,\ell)}{S_2^O(\beta,\ell)}\right) \qquad (4.10)$$

where S_2^P and S_2^O represent the SFs respectively with and without perturbation. It characterizes the relative disturbance in dB. Figure 4.4b shows the measured $\mathcal{I}(\beta,\ell)$ for the same β as shown in Figure 4.4a. Note that with a large value of β, the influence of a single scale is more visible. In reality, the inertial range is always finite due to the finite size of the system. Therefore, the convergence problem is not an issue for the application of Equation (4.3). Here we calculate the second-order SFs by applying the Equation (4.3) even for a value of β out of the convergence

4.1 Structure function analysis 61

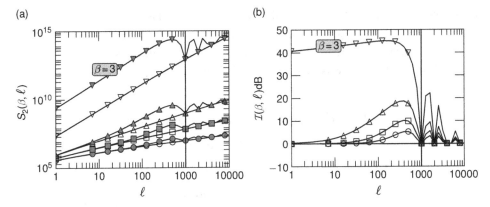

Figure 4.4 a) Measured second-order structure-function without (open symbols) and with perturbation (solid symbols) for the same β as in Figure 4.3. b) The influence function $\mathcal{I}(\beta,\ell)$ in dB. Note that the larger the β value, the more $S_2(\ell)$ or $\mathcal{I}(\beta,\ell)$ are influenced by the perturbation.

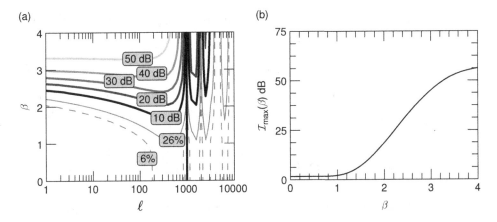

Figure 4.5 Contour plot of the measured influence function $\mathcal{I}(\beta,\ell)$ in dB.

range $[1,3]$. Figure 4.5 a) shows a contour plot of the measured $\mathcal{I}(\beta,\ell)$, in which contour lines for 0.5 and 2 dB (resp. 6% and 26% disturbance) are illustrated by dashed and solid lines. The maximum disturbance is defined as:

$$\mathcal{I}_{\max}(\beta) = \max_{\ell}\{\mathcal{I}(\beta,\ell)\} \qquad (4.11)$$

Figure 4.5 b) shows the measured value of $\mathcal{I}_{\max}(\beta)$. One can see that the measured value of $\mathcal{I}_{\max}(\beta)$ increases with β, confirming again that the large-scale influence is large for large value of β, the slope of the Fourier power spectrum.

The above discussion shows that the SF might be influenced by large scale structures, such as annual cycle present in many collected fluid geosciences data.

Therefore, the application of the SF analysis to real data should be done with caution. Several methods have been developed to constrain the large scale influence. These are the detrended fluctuation analysis, detrended structure function, and Hilbert spectral analysis, to name a few. They are described below.

4.1.4 Generalization for high-order moments

A natural generalization of the second-order structure function Equation (4.1) for arbitrary q is written as:

$$S_q(\ell) = \langle |u(x+\ell) - u(x)|^q \rangle \qquad (4.12)$$

in which $q \geq -1$. For a scaling process, we expect the following power-law behavior:

$$S_q(\ell) \sim \ell^{\zeta(q)} \qquad (4.13)$$

in which $\zeta(q)$ is the scaling exponent. As detailed in Chapter 3, for a fBm process, the corresponding scaling exponent is written as:

$$\zeta(q) = qH \qquad (4.14)$$

in which H is a Hurst number. The measured $\zeta(q)$ is linear when the analyzed process is monofractal and is usually convex when the analyzed process is multifractal (Frisch, 1995).

Figure 4.6 shows the measured structure-function $S_q(\ell)$ for the fBm process with a Hurst number $H = 1/3$. Power-law behavior is observed on all separation scales and all q considered here. The solid line illustrates the power-law fitting. The corresponding scaling exponent $\zeta(q)$ is shown as inset, in which the theoretical prediction qH is represented by a solid line. A 95% fitting confidence limit is indicated by the errorbar, which is within the symbol size. The nice agreement between the measured scaling exponent and the theoretical prediction shows the efficiency of the structure function analysis for this simple monofractal process (Huang et al., 2010).

We provide here some comments on the structure function analysis. As we mentioned above, it is strongly influenced by the large-scale structure, known as infrared effect, especially when $\beta \geq 2$ (Huang et al., 2011a, 2013). Conversely, it is also strongly contaminated by the energetic small-scale structures, such as vortex trapping in Lagrangian turbulence. This small-scale contamination is now recognized as an ultraviolet effect (Huang et al., 2013). Therefore, the extracted high-order scaling exponent $\zeta(q)$ might be significantly biased due to these two effects. An ideal approach should avoid the scale mixing problem and should constrain both the infrared and ultraviolet effects.

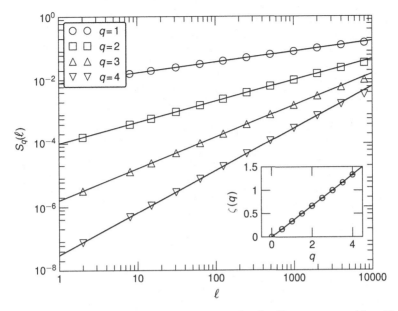

Figure 4.6 Measured structure-function $S_q(\ell)$ for the fBm process with a Hurst number $H = 1/3$ and a data length 10^6 points. Power-law behavior is observed for all q considered here. The inset shows the measured scaling exponent $\zeta(q)$, in which the theoretical value qH is illustrated by a solid line. The errorbar is a 95% fitting confidence limit.

4.2 Autocorrelation function of increments

4.2.1 Experimental observation

In Anselmet et al. (1984) classical paper, the authors found experimentally (without theoretical explanation) that the autocorrelation function of increments has its minimum value at the location $\tau = \ell$, in which ℓ is the separation scale of the increment and τ is the lag for the autocorrelation function. The autocorrelation function of the increment is defined as:

$$\Gamma_\ell(\tau) = \langle \Delta_\ell x(t) \Delta_\ell x(t+\tau) \rangle_t \tag{4.15}$$

in which $\Delta_\ell x = x(t+\ell) - x(t)$ is the increment with a scale ℓ. Their finding is written as:

$$\Gamma_{\min}(\ell) = \min_\tau \{\Gamma_\ell(\tau)\} \tag{4.16a}$$

and τ_0 the location of the minimum

$$\Gamma_{\min}(\ell) = \Gamma_\ell(\tau_0(\ell)), \quad \tau_0(\ell) = \ell \tag{4.16b}$$

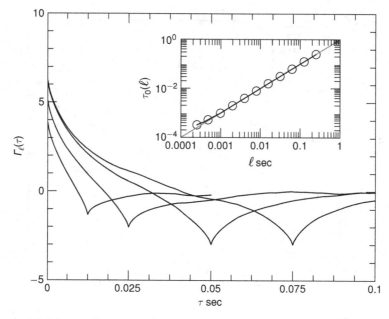

Figure 4.7 Measured autocorrelation function $\Gamma_\ell(\tau)$ of velocity increment with various separation scale ℓ for a turbulent velocity. The turbulent velocity is obtained from a high Reynolds number wind tunnel experiment with $Re_\lambda = 720$. Inset shows the measured location $\tau_0(\ell)$ for the minimum value of $\Gamma_\ell(\tau)$, in which the solid line indicates the value $\tau_0(\tau) = \ell$. The collapse of the symbols and solid line verifies Equation (4.16b) on a large range of scales.

Figure 4.7 shows the measured $\Gamma_\ell(\tau)$ for velocity increments with various separation scale ℓ. The turbulent velocity is obtained from a high Reynolds number wind tunnel experiment with $Re_\lambda = 720$. A minimum value $\Gamma_{\min}(\ell)$ is observed for all separation scale ℓ. The corresponding location $\tau_0(\ell)$ is displayed as inset. Visually, it confirms the Equation (4.16b) on a large range of scales. Below we prove Equation (4.16b) analytically for a scaling process.

4.2.2 Theoretical demonstration

Let us consider now the autocorrelation function of the increment $\Delta_\ell x$. The Wiener-Khinchin theorem relates the autocorrelation function to the power spectral density via the Fourier transform (Percival and Walden, 1993; Frisch, 1995)

$$\Gamma_\ell(\tau) = \int_0^{+\infty} E_\Delta(f) \cos(2\pi f \tau) df \qquad (4.17)$$

in which $E_\Delta(f)$ is the Fourier power spectrum of the increment $\Delta_\ell x(t)$. The theorem can be applied to wide-sense-stationary random processes, signals whose Fourier

4.2 Autocorrelation function of increments

transform may not exist, using the definition of autocorrelation function in terms of expected value rather than an infinite integral (Percival and Walden, 1993). Substituting Equation (4.3) ($E_\Delta(f) = E(f)(1 - \cos(2\pi f \ell))$) into the above equation, and assuming a power law for the spectrum (a hypothesis of similarity)

$$E(f) = cf^{-\beta}, \quad c > 0 \tag{4.18}$$

we obtain

$$\Gamma_\ell(\tau) = c \int_0^{+\infty} f^{-\beta}(1 - \cos(2\pi f \ell)) \cos(2\pi f \tau) df \tag{4.19}$$

The convergence condition requires $0 < \beta < 3$. It implies a rescaled relation, using scaling transformation inside the integral. This can be estimated by considering $\Gamma_{\lambda\ell}(\tau)$ and taking $f' = f\lambda$ for $\lambda > 0$, providing the identity (after adequate rescaling into the integral)

$$\Gamma_{\lambda\ell}(\tau) = \Gamma_\ell(\tau/\lambda)\lambda^{\beta-1} \tag{4.20}$$

If we take $\ell = 1$ and replace λ by ℓ, we then have

$$\Gamma_\ell(\tau) = \Gamma_1(\tau/\ell)\ell^{\beta-1} \tag{4.21}$$

Thus, we have a universal autocorrelation function

$$\Gamma_\ell(\ell\varsigma)\ell^{1-\beta} = \Upsilon(\varsigma) = \Gamma_1(\varsigma) \tag{4.22}$$

This rescaled universal autocorrelation function is shown in Figure 4.8 for various scales ℓ, verifying the above scaling relation experimentally. A derivative of Equation (4.20) gives $\Gamma'_{\lambda\ell}(\tau) = \Gamma'_\ell(\tau/\lambda)\lambda^{\beta-2}$. The minimum value of the left-hand side is $\tau = \tau_o(\lambda\ell)$, verifying $\Gamma'_{\lambda\ell}(\tau_o(\lambda\ell)) = 0$, and for this value we have also $\Gamma'_\ell(\tau_o(\lambda\ell)/\lambda) = 0$. This shows that $\tau_o(\ell) = \tau_o(\lambda\ell)/\lambda$. Taking again $\ell = 1$ and $\lambda = \ell$, we have

$$\tau_o(\ell) = \ell\tau_o(1) \tag{4.23}$$

Showing that $\tau_o(\ell)$ is proportional to ℓ in the scaling range (when ℓ belongs to the inertial range). With the definition of $\Gamma_o(\ell) = \Gamma_\ell(\tau_o(\ell))$ we have, also using Equation (4.21), for $\tau = \tau_o(\lambda\ell)$:

$$\begin{aligned}\Gamma_{\lambda\ell}(\tau_o(\lambda\ell)) &= \Gamma_\ell(\tau_o(\lambda\ell)/\lambda)\lambda^{\beta-1} \\ &= \Gamma_\ell(\tau_o(\ell))\lambda^{\beta-1}\end{aligned} \tag{4.24}$$

Hence $\Gamma_o(\lambda\ell) = \lambda^{\beta-1}\Gamma_o(\ell)$ or

$$\Gamma_o(\ell) = \Gamma_o(1)\ell^{\beta-1} \tag{4.25}$$

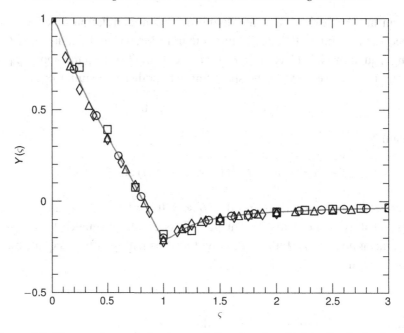

Figure 4.8 Measured rescaled autocorrelation function $\Gamma_\ell(\tau)$ of velocity increments with various separation scale ℓ. The collapse of symbols confirm the universal rescaling Equation (4.22).

We now consider the location $\tau_o(1)$ of the autocorrelation function for $\ell = 1$. For this, we note:

$$I = \int_0^\infty u^{-\beta} \cos(2\pi u) du \qquad (4.26)$$

and we develop the cosine product inside the integral definition of $\Gamma_\ell(\tau)$. We obtain after performing adequate rescaling into the obtained integrals, the following expression:

$$\Gamma_\ell(\tau) = \left(\tau^{\beta-1} - \frac{1}{2}(\ell - \tau)^{\beta-1} - \frac{1}{2}(\ell + \tau)^{\beta-1} \right) I \qquad (4.27)$$

which verifies indeed, as shown above, $\Gamma_\ell(\tau/\ell) = \ell^{1-\beta}\Gamma_\ell(\tau)$. To find the minimum value of $\Gamma_1(\tau)$, let us note: $\mathcal{P}(\tau) = \frac{d\Gamma_1(\tau)}{d\tau}$. We obtain

$$\frac{2\mathcal{P}(\tau)}{(\beta - 1)I} = 2\tau^{\beta-2} + (1 - \tau)^{\beta-2} - (1 + \tau)^{\beta-2} \qquad (4.28)$$

The convergence condition requires $1 < \beta < 4$. When $\beta < 2$, one can find that both left and right limits of $\mathcal{P}(1)$ are infinite, but the definition of $\mathcal{P}(1)$ is finite. Thus $\tau = 1$ is a second type discontinuity point of Equation (4.27) (Malik and Arora, 1992). It is easy to show that

$$\begin{cases} \mathcal{P}(\tau) < 0, \tau \le 1 \\ \mathcal{P}(\tau) > 0, \tau > 1 \end{cases} \qquad (4.29)$$

It means that $\mathcal{P}(\tau)$ changes its sign from negative to positive when τ is increasing from $\tau < 1$ to $\tau > 1$. In other words, the autocorrelation function will take its minimum value at the location where τ is exactly equal to 1. We thus see that $\tau_o(1) = 1$ and hence $\tau_o(\ell) = \ell$ (Equation [4.16b]). Thus we have shown here analytically that for a scaling time series with $1 < \beta < 2$ the relation found experimentally by Anselmet et al. (1984) is true. This was first published in Huang et al. (2011b).

4.2.3 Power-law behavior

The above analysis provides a clear scaling behavior of the autocorrelation function of increments (also see Equations [4.21], [4.22] and [4.25]),

$$\Gamma_o(\ell) \sim \ell^{\beta-1} \qquad (4.30)$$

in which β is the slope of the Fourier power spectrum. Figure 4.9 shows the measured $\Gamma_o(\ell)$ for a synthesized fBm ($\square$) with a Hurst number $H = 1/3$ and 10^6 data points, and for a turbulent velocity ($\bigcirc$), respectively. Power-law behavior is observed for both data sets with scaling exponents 0.67 ± 0.01 and 0.80 ± 0.02, verifying the scaling behavior prediction by Equation (4.30) (Huang et al., 2009b). To emphasize the observed power-law, the inset shows the compensated curve $\Gamma_o(\ell)\ell^{-0.8}C^{-1}$, in which C is a prefactor from the least square fitting algorithm. A clear plateau is observed for nearly two decades. For comparison, the compensated curve for the second-order structure function is also shown as a solid line. Visually, the plateau provided by the second-order structure function is much shorter than the one provided by the $\Gamma_o(\ell)$. This phenomenon has been interpreted as the influence of large scale structures (Huang et al., 2009b, 2010).

4.2.4 Cumulative function

Similarly, as with the second-order structure function, we shall introduce the cumulative function as below:

$$\mathcal{Q}(f, \ell) = \frac{\int_0^f E(f')\big(1 - \cos(2\pi f'\ell)\big) \cos(2\pi f'\ell) df'}{\int_0^\infty E(f')\big(1 - \cos(2\pi f'\ell)\big) \cos(2\pi f'\ell) df'} \times 100\% \qquad (4.31)$$

in which $E(f)$ is the Fourier power spectrum of the original variable. This characterizes the relative contribution of the frequency band $[0, f]$: when $\mathcal{Q}(f, \ell)$ is small, the relative contribution of the frequency band $[0, f]$ to the denominator expression

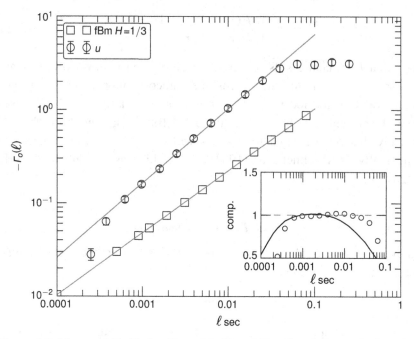

Figure 4.9 Measured $\Gamma_o(\ell)$ for fBm with $H = 1/3$ and turbulent velocity with $Re_\lambda = 720$. Specifically for the turbulent velocity, a power-law behavior with a scaling exponent 0.80 ± 0.02 is observed on the range $5 \times 10^{-4} < \ell < 5 \times 10^{-2}$ sec, corresponding to a frequency range $200 < f < 2000$ Hz. The error bar is a standard deviation from 120 realizations. The inset shows the compensated curve to emphasize the observed power-law behavior. For comparison, the compensated curve for the second-order structure function is also shown as a solid line. Note that the plateau for the structure function is much shorter than the one for $\Gamma_o(\ell)$. This phenomenon has been interpreted as the large-scale influence.

is small. Assuming a pure power-law Fourier spectrum $E(f) \sim f^{-\beta}$, one obtains an analytical expression for Equation (4.31),

$$Q(f, \ell) = \frac{1}{b(\beta)} \left\{ (3-\beta) [\cos(f) - 1] \cos(f) f^{1-\beta} + g(f, \beta) f^{3-\beta} \right\} \times 100\% \quad (4.32)$$

in which $b(\beta) = -\sqrt{\pi}(3-\beta)(2^{1-\beta} - 1/2)\Gamma(3/2 - \beta/2)\Gamma(\beta/2)^{-1}$ and $g(f, \beta) = {}_2{}_1F_2(3/2 - \beta/2, 3/2, 5/2 - \beta/2, -f^2) - {}_1F_2(3/2 - \beta/2, 5/2 - \beta/2, -f^2/4)$ and ${}_1F_2$ is a generalized hypergeometric function. Note that $Q(f, \ell)$ is independent of the separation scale ℓ.

Figure 4.10 shows the measured $Q(f, 1)$ (solid line) with $\beta = 5/3$. It is first negative and decreases with f until $f \simeq 0.25$. It then increases with f until $f \simeq 0.75$ with $Q(0.75, 1) \simeq 0.53$. It then approaches to 100% with an oscillation. Note that the measured $Q(f, 1)$ crosses zero when $f \simeq 0.49$. This implies that the contribution from the frequency band $[0, 0.49]$ is canceled by themselve. In other words, the influence of large scale structures is constrained (Huang et al., 2010). This is the

4.3 Maximum probability density function scaling of velocity increments

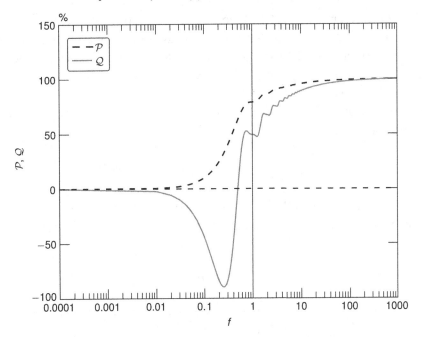

Figure 4.10 Cumulative function $\mathcal{P}(f,1)$ (dashed line) for the second-order structure function and $\mathcal{Q}(f,1)$ for autocorrelation function of increments (solid line). Note that $\mathcal{Q}(0.496,1) \simeq 0$, indicating that the influence of the large scale structure might be canceled by each other for $\Gamma_o(\ell)$.

reason why $\Gamma_o(\ell)$ provides a better scaling range than the structure function. For comparison, the measured $\mathcal{P}(f,1)$ for the second-order structure function is also reproduced as a dashed line. It is found that the contribution from the large scale part ($f < 1$) represents 49.01% and 78.76% respectively for $\mathcal{Q}(f,1)$ and $\mathcal{P}(f,1)$. Hence large scales have much more influence for structure function than for the $\Gamma_o(\ell)$ method.

4.3 Maximum probability density function scaling of velocity increments

Recently, Huang et al. (2008a) found a Hilbert-based probability density function (pdf) scaling for turbulent velocity. Inspired by these findings, Huang et al. (2011b) discovered an increment-based pdf scaling. Below we provide three different ways to consider this pdf scaling.

4.3.1 Fractional Brownian motion

We shall first consider here an fBm process. As we have already mentioned in Chapter 3, the fBm is a continuous-time random process proposed by Kolmogorov (1940) and Yaglom (1957) and later named "fractional Brownian motion" by

Mandelbrot and Van Ness (1968). It consists in a fractional integration of a white Gaussian process and is therefore a generalization of Brownian motion, which consists simply in a standard integration of a white Gaussian process. Below we shall consider it as an analytical model for monofractal processes to obtain the pdf scaling analytically.

An autocorrelation function of fBm's increments $Y_\tau(t) = x(t+\tau) - x(t)$ is known to be the following:

$$R_\tau(\ell) = \frac{1}{2}\{(\tau + \ell)^{2H} + |\tau - \ell|^{2H} - \ell^{2H}\} \tag{4.33}$$

where $\ell \geq 0$ is the time delay, τ is the separation scale, and H is Hurst number (Biagini et al., 2008). Thus the standard deviation $\sigma(Y_\tau)$ of the increment Y_τ scales as:

$$\sigma(Y_\tau) = R_\tau(0)^{1/2} = \tau^H \tag{4.34}$$

Y_τ is also known to have a Gaussian distribution (Mandelbrot and Van Ness, 1968; Biagini et al., 2008) which reads as

$$p(Y_\tau) = \frac{1}{\sigma(Y_\tau)\sqrt{2\pi}} \exp\left(-\frac{Y_\tau^2}{2\sigma(Y_\tau)^2}\right) \tag{4.35}$$

We therefore have a power law relation when $Y_\tau = 0$

$$p_{\max}(\tau) = p(Y_\tau)|_{Y_\tau=0} = \frac{1}{\sigma(Y_\tau)\sqrt{2\pi}} = \frac{1}{\sqrt{2\pi}} \tau^{-\alpha(H)} \tag{4.36}$$

where $\alpha(H) = H$. Hence, we obtain a pdf scaling for the fBm process:

$$p_{\max}(\tau) \sim \tau^{-H} \tag{4.37}$$

in which H is the Hurst number.

4.3.2 H-self-similarity processes

We can also derive the pdf scaling more generally for H-self-similar processes as follows. We define a H-self-similar process as (see Chapter 3)

$$\{x(at)\} \stackrel{d}{=} \{a^H x(t)\} \tag{4.38}$$

in which $\stackrel{d}{=}$ means equality in distribution and H is the Hurst number (Embrechts and Maejima, 2002; Samorodnitsky and Taqqu, 1994). $x(t)$ is a H-self-similar

4.3 Maximum probability density function scaling of velocity increments

process, in which only one parameter H, namely the Hurst number, is required for the above scaling transform. Let us note $Y_\tau = \Delta x_\tau = x(t+\tau) - x(t)$, the increment with separation scale τ. We assume that x is H-self-similar with stationary increment, hence Y_τ is also H-self-similar. Thus one has

$$\left\{\frac{Y_\tau}{\tau^H}\right\} \stackrel{d}{=} \left\{\frac{Y_T}{T^H}\right\} \stackrel{d}{=} \{Y_1\} \qquad (4.39)$$

In fact equality in distribution means equality for distribution function. Let us write the distribution function

$$F(x) = P_r(X \leq x) = \int_{-\infty}^{x} p(u)du \qquad (4.40)$$

in which $p(x)$ is the pdf of x. We note the pdf

$$p(x) = F'(x) \qquad (4.41)$$

We accordingly take here

$$F_\tau(x) = P_r(Y_\tau \leq x) \qquad (4.42)$$

We have

$$P_r\left(\frac{Y_\tau}{\tau^H} \leq x\right) = P_r\left(Y_\tau \leq x\tau^H\right) \qquad (4.43)$$

Hence, Equation (4.39) writes for distribution functions

$$F_\tau\left(x\tau^H\right) = F_T\left(xT^H\right) \qquad (4.44)$$

Taking the derivative of Equation (4.44), we have for the pdfs

$$\tau^H p_\tau\left(x\tau^H\right) = T^H p_T\left(xT^H\right) \qquad (4.45)$$

Then writing

$$p_{\max}(\tau) = \max_x \{p_\tau(x)\} \qquad (4.46)$$

and taking the maximum of Equation (4.45), we have

$$\tau^H p_{\max}(\tau) = T^H p_{\max}(T) \qquad (4.47)$$

Finally, this leads to

$$p_{\max}(\tau) = p_{\max}(T)(\tau/T)^{-H} \qquad (4.48)$$

This is the pdf scaling for the H-self-similar process.

Since Equation (4.38) is not true for multi-scaling processes, Equation (4.48) may be only an approximation for multifractal processes.

4.3.3 Dimensional analysis for increments

The pdf scaling can be obtained also by using the dimensional argument as shown in Qiu et al. (2014). Taking a turbulent velocity time series $u(t)$ as example, the velocity increment is defined as usual

$$X_\tau(t) = \Delta u_\tau(t) = u(t+\tau) - u(t) \qquad (4.49)$$

in which τ is the separation time. One can define a pdf $p_\tau(X_\tau)$ of $X_\tau(t)$ for a given separation time τ. Note that the probability $p_\tau(X_\tau)dX_\tau$ is a pure number without dimension. Thus one has

$$[p_\tau(X_\tau)] = [dX_\tau]^{-1} = [X_\tau]^{-1} = [u]^{-1} \qquad (4.50)$$

in which $[\cdot]$ means dimension. According to the Kolmogorov's similarity hypotheses (Kolmogorov, 1941b; Frisch, 1995) we therefore expect the following power law

$$p_\tau(X_\tau) \sim X_\tau^{-1} \sim \tau^{-\alpha} \qquad (4.51)$$

in which α is comparable with the first-order SF scaling exponent $\zeta(1)$ and is to be $1/3$ for the Kolmogorov non intermittent value of homogeneous and isotropic turbulent flows (Huang, 2009; Huang et al., 2011b). However, the above Equation (4.51) does not hold for the whole pdf $p_\tau(X_\tau)$. As already shown above the maximum pdf $\max_{X_\tau}\{p_\tau(X_\tau)\}$ at a given scale can be considered as a surrogate of the whole pdf $p_\tau(X_\tau)$,

$$p_{\max}(\tau) = \max_{X_\tau}\{p(X_\tau)\} \sim \tau^{-\alpha} \qquad (4.52)$$

As mentioned above, the scaling exponent α is comparable with the first-order SF scaling exponent $\zeta(1)$ (Huang et al., 2011b).

Figure 4.11 shows the measured $p_\tau(x)$ for two separation scales. Experimentally, the maximum pdf of the increment $p_\tau(x)$ is usually taken around $x \simeq 0$. Therefore, $p_{\max}(\tau)$ is interpreted as background fluctuations of the given fields, in which the effect of large scale structures, e.g., thermal plumes in the Rayleigh-Bénard system, is excluded (Huang et al., 2011b; Qiu et al., 2014). Note that this method might require a larger data sample size than the classical structure function since only background fluctuation of the given process are considered. Figure 4.12 shows the measured $p_{\max}(\tau)$ for the fBm process with a Hurst number $H = 0.5$ and 100 realizations with $L = 10^6$ data point each. The errorbar indicates the standard deviation from these 100 realizations. The corresponding scaling exponent is found to be $\alpha = 0.50 \pm 0.01$. To test the finite size effect, we also calculate $p_{\max}(\tau)$ for different sample sizes L. The inset of Figure 4.12 displays the measured L-dependent scaling exponent $\alpha(L)$. Visually, the experimental $\alpha(L)$ is slightly

4.3 Maximum probability density function scaling of velocity increments

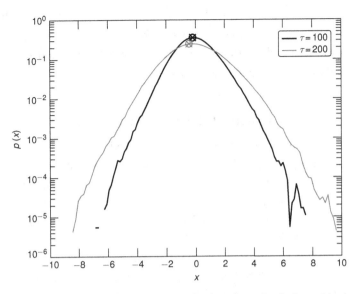

Figure 4.11 Experimental probability density function of turbulent velocity increments. The turbulent velocity is obtained from a high Reynolds wind tunnel with $Re_\lambda = 720$. The separation τ is in data points. Note that the location of the maximum value of $p(x)$ is close to zero.

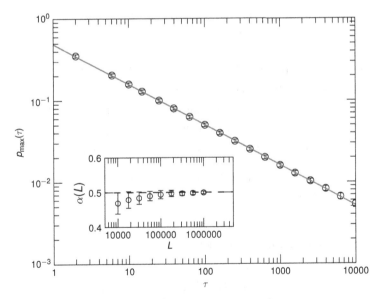

Figure 4.12 Measured $p_{\max}(\tau)$ for fBm process with a Hurst number $H = 0.5$ and 100 realizations and $L = 10^6$ data points each. The errorbar indicates the standard deviation from these 100 realizations. The measured scaling exponent is $\alpha = 0.50 \pm 0.01$. The inset shows the finite sample size effect obtained from 100 realizations with different data length L. Note that with the increasing of the data length L, the measured α is approaching the given value of $\alpha = 0.5$.

below $H = 0.5$. It is increasing with L and approaching to $H = 0.5$. It is found that with $L = 10^5$, the relative error between the measured α and given H is less than 1%. For the first-order structure function, this error is found to be less than 0.1% (Huang et al., 2011b; Qiu et al., 2014).

4.4 Detrended fluctuation analysis

Detrended Fluctuation Analysis (DFA) was first introduced by Peng et al. (1994) to study the scaling properties of the DNA sequence, in which only the second-order moment $q = 2$ was considered. Later it was generalized into a multifractal version by considering the arbitrary order q, namely multifractal detrended fluctuation analysis (MFDFA) (Kantelhardt et al., 2002; Oświęcimka et al., 2006). It then became a rather common technique for scaling data analysis (Peng et al., 1994; Heneghan and McDarby, 2000; Hu et al., 2001; Kantelhardt et al., 2002; Chen et al., 2002; Koscielny-Bunde et al., 2006; Sadegh Movahed et al., 2006; Oświęcimka et al., 2006; Bardet and Kammoun, 2008; Zhang et al., 2008; Bashan et al., 2008). For a given discrete time series $x(i)$, $i = 1 \cdots N$, we shall first estimate its cumulative function

$$Y(j) = \sum_{i=1}^{j} (x(i) - \bar{x}), \quad j = 1, \cdots N \tag{4.53}$$

where $\bar{x}$ is the mean value of x. We then divide $Y(j)$ into M_n segments of length n ($n < N$) starting from both the beginning and the end of the time series. Each segment v has its own local trend that can be approximated by fitting a pth-order polynomial P_v^p, which is then removed from the data (see Figure 4.13). The variances for all the segments v and for all segment lengths n are then calculated by

$$F^2(v, n) = \frac{1}{n} \sum_{j=1}^{n} \{Y[(v-1)n + j] - P_v^p(j)\}^2 \tag{4.54}$$

The qth-order fluctuation function is then defined as

$$F^q(n) = \left(\frac{1}{2M_n} \sum_{v=1}^{2M_n} [F^2(v, n)]^{q/2} \right)^{1/q} \tag{4.55}$$

For discussion convenience, we redefine the qth-order fluctuation function as

$$\mathcal{F}_q(n) = F^q(n)^q \tag{4.56}$$

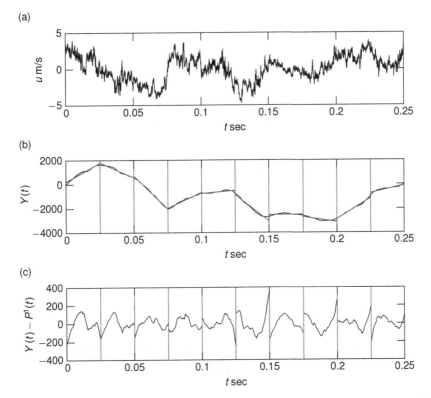

Figure 4.13 An illustration of the DFA procedure. a) Measured velocity $u(t)$ with a portion of length 0.25 sec. The experimental data are obtained from a high Reynolds number wind tunnel experiment (Kang et al., 2003). b) The measured cumulative function $Y(t)$, in which the thin line is the first-order trend with 0.025 sec. c) The first-order detrended data $Y(t) - P^1(t)$, showing the detail after the detrending.

In case of scale invariance, we have power law scaling within a significant range of n

$$\mathcal{F}_q(n) \sim n^{h(q)} \qquad (4.57)$$

in which $h(q)$ is the corresponding scaling exponent function. Note that we have $F^2(v, n) > 0$, so we can retrieve $h(q)$ for both positive and negative q. For an fBm process, the measured scaling exponents $h(q)$ can be written as,

$$h(q) = (1 + H)q \qquad (4.58)$$

in which H is the Hurst number. Note that for the structure function, the corresponding scaling exponent is $\zeta(q) = qH$ (see Equation [4.14]). The difference between Equations (4.14) and (4.58) comes from the fact that during the DFA procedure, the cumulative function, Equation (4.53), is introduced.

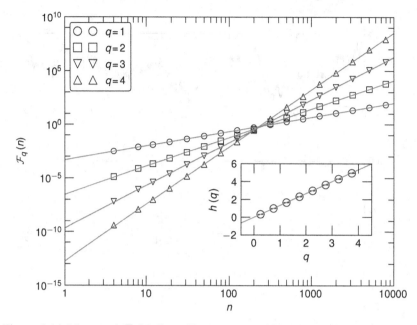

Figure 4.14 Measured $\mathcal{F}_q(n)$ for a fBm process with a Hurst number $H = 1/3$, in which the solid line is a power-law fit. The inset shows the measured scaling exponent $h(q)$, in which the solid line is the theoretical value $h(q) = (1 + H)q$. The errorbar indicates the 95% fitting confidence limit.

Figure 4.14 shows the measured $\mathcal{F}_q(n)$ for the fBm process with $H = 1/3$ and a data length of 10^6 data point. Power-law behavior is observed as expected, confirming the validation of the scaling relation of Equation (4.57). The corresponding scaling exponent $h(q)$ is shown as inset, in which the 95% confidence limit is illustrated by an errorbar. The theoretical value provided by Equation (4.58) is demonstrated by a solid line. Visually, the DFA provides a nice measurement of $h(q)$.

Let us comment here on this method. The DFA is proposed to handle non-stationary events, in which a detrending method is employed. For this method, the shortcoming of the classical structure function, such as large-scale influence (infrared effect) and energetic small-scale contamination (ultraviolet effect) are expected to be constrained.

However, in practice, it is found that this method is still influenced by the above mentioned effects, namely scale-mixing problem (Huang et al., 2011a). This is partially because the cumulative function (Equation [4.53]) mixes the information from different scales. It is also partially because the detrending method cannot remove the trend accurately, since the trend could be linear or nonlinear with different shapes. For example, in the passive scalar, due to the existence of ramp-cliff structures, the DFA is strongly influenced by such structures (Huang et al., 2011a).

Therefore, different detrending approaches may be employed with different performances: for example, the moving average, or Empirical Mode Decomposition, to mention a few. Different detrending methods might have different performances. In practice, we have found that in most situations, the first-order DFA already provides a good performance.

Let us note that the DFA method has been quite popular in the last years, applied in many fields. However we feel such popularity is overdue, since the method does not perform much better than structure functions. In most cases when it is applied, the classical structure functions could also be applied; and when structure functions is not applicable, the DFA method is not the best alternative.

4.5 Detrended structure function

As mentioned above the classical structure function is influenced by energetic structures; known as infrared effect for the large-scale structure effect and ultraviolet effect for the small-scale structure effect. The detrended fluctuation analysis attempts to constrain the nonstationary effect by introducing a detrending process. Below we have introduced a detrended structure function analysis, in which the detrending analysis is performed before applying the standard structure function analysis. By doing so, the large-scale structure effect would then be expected to be constrained.

4.5.1 Detrending analysis

We start here with a scaling process $x(t)$, which possesses a power law Fourier spectrum,

$$E(f) = Cf^{-\beta} \tag{4.59}$$

in which β is the scaling exponent. Parseval's theorem states the following relation (Percival and Walden, 1993):

$$\langle x(t)^2 \rangle_t = \int_{-\infty}^{+\infty} E(f) df \tag{4.60}$$

in which $E(f)$ is the Fourier power spectrum of $x(t)$. We first divide the given $x(t)$ into n segments with a length τ each. A qth-order detrending of the ith segment is defined as:

$$x_{i,\tau}(t) = x_i(t) - P_{i,\tau}^q(t), \ (i-1)\tau \leq t \leq i\tau \tag{4.61}$$

in which $P_{i,\tau}^q(t)$ is a qth-order polynomial fitting of the $x_i(t)$. We denote this as detrending analysis.

To obtain a detrended signal, i.e., $x_\tau(t) = [x_{1,\tau}(t), x_{2,\tau}(t) \cdots x_{n,\tau}(t)]$, a linear trend (resp. 1st-order polynomial fitting) or a local mean (resp. zero-order polynomial fitting) is removed within a window size τ. Ideally, the scales larger than τ, i.e., $r > \tau$ are removed or constrained from the original data $x(t)$. The kinetic energy of $x_\tau(t)$ could be related directly with its Fourier power spectrum, i.e.,

$$\mathcal{D}_2(\tau) = \langle x_\tau(t)^2 \rangle_t = \int_{-\infty}^{+\infty} E_\tau(f) df \simeq \int_{f_\tau}^{+\infty} E(f) df \quad (4.62)$$

in which $f_\tau = 1/\tau$. This implies that the detrending procedure acts as a high-pass filter, in which the lower Fourier modes $f < f_\tau$ (resp. $r > \tau$) are expected to be removed or constrained. For a scaling process, i.e., $E(f) \sim f^{-\beta}$, this yields a power-law behavior, i.e.,

$$\mathcal{D}_2(\tau) \sim f_\tau^{1-\beta} \sim \tau^{\beta-1} \quad (4.63)$$

The convergence condition requires $\beta \geq 1$. The physical meaning of $\mathcal{D}_2(\tau)$ is quite clear: it represents a cumulative energy over the Fourier frequency band $[f_\tau, +\infty]$ (Huang, 2014).

Figure 4.15 a shows the original time series $x(t)$ for the fBm process with a Hurst number $H = 1/3$, in which the linear trend with a scale $\tau = 1000$ data points is illustrated by a thick solid line. The corresponding detrended time series $x_\tau(t)$ is shown in Figure 4.15 b. Visually, large-scale fluctuations $r \geq \tau$ are removed. To emphasize this high-pass filter property, we calculate the Fourier power spectrum $E_\tau(f)$ for several scales τ. Figure 4.16 shows the measured Fourier spectrum $E_\tau(f)$, in which the separation scale $f = 1/\tau$ is illustrated by a vertical line. It confirms that the large-scale fluctuations $r \geq \tau$ are constrained. Figure 4.17 shows the measured $\mathcal{D}_2(\tau)$, in which the inset shows the compensated curve with a scaling exponent 2/3 as predicted by Equation (4.63). Power-law behavior is observed on the range $10 < \tau < 10000$, verifying the prediction of Equation (4.63).

4.5.2 Detrended structure function

After the detrending procedure of the ith segment, an increment is then defined within the window size τ as:

$$\Delta x_{i,\tau}(t) = |x_{i,\tau}(t + \tau/2) - x_{i,\tau}(t)| \quad (4.64)$$

We will show in the next subsection why an increment with a half width of the window size $\tau/2$ is introduced. An nth-order detrended structure function (DSF) is then defined as:

$$\mathcal{B}_q(\tau) = \langle \Delta x_{i,\tau}(t)^q \rangle_t \quad (4.65)$$

4.5 Detrended structure function

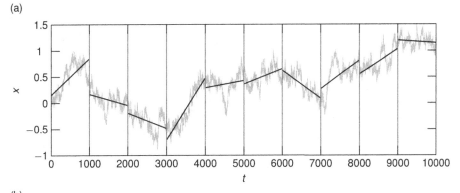

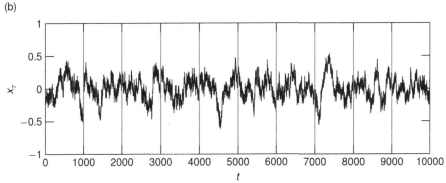

Figure 4.15 Illustration of the first-order detrending analysis for the fBm process with $H = 1/3$ and data length 10^4 points. a) The original time series (grey line) and the linear trend with $\tau = 1000$ data points. b) The detrended time series x_τ. Ideally, the large-scale fluctuation $r \geq \tau$ would then be expected to be removed.

For a scaling process, we expect a power behavior of DSFs:

$$\mathcal{B}_q(\tau) \sim \tau^{\zeta(q)} \quad (4.66)$$

in which the scaling exponent $\zeta(q)$ is comparable with the one provided by the original SFs.

To access negative orders of n (the right part of the singularity spectrum $D(h)$), the DSFs can be redefined as:

$$\mathcal{B}_q(\tau) = \langle |X_\tau(i)|^q \rangle \quad (4.67)$$

in which $X_\tau(i) = \langle |\Delta x_{i,\tau}(t)| \rangle_{(i-1)\tau \leq t \leq i\tau}$ is local average for the ith segment. A power-law behavior is also expected, i.e., $\mathcal{B}_q(\tau) \sim \tau^{\zeta(q)}$. It is found that when $q \geq 0$, Equations (4.65) and (4.67) provide the same scaling exponents $\zeta(q)$.

Figures 4.18 a and b show the measured $\mathcal{B}_q(\tau)$ provided by Equation (4.65) and (4.67) for the fBm process with a Hurst number $H = 1/3$. Power-law behavior

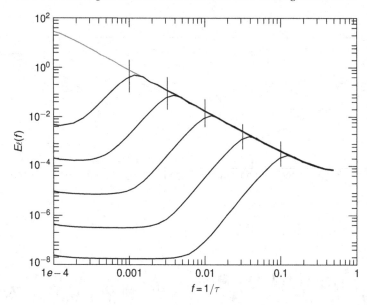

Figure 4.16 Comparison of the Fourier power spectrum for original time series $x(t)$ and the detrended time series x_τ for various detrending scales τ, in which the vertical solid lines indicate the scale $f = 1/\tau$. Visually the large-scale fluctuations $f < 1/\tau$ are constrained, verifying Equation (4.62).

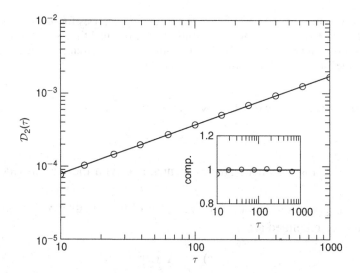

Figure 4.17 Experimental $\mathcal{D}_2(\tau)$ for the fBm process with a Hurst number $H = 1/3$. Power-law behavior is observed as predicted by Equation (4.63). The inset shows the compensated curve with a theoretic scaling exponent 2/3.

4.5 Detrended structure function

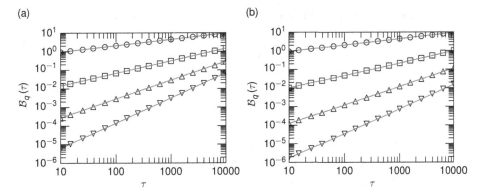

Figure 4.18 Experimental $\mathcal{B}_q(\tau)$ for the fBm process with a Hurst number $H = 1/3$ and a data length 10^6 data point: a) Equation (4.65) and b) Equation (4.67). Power-law behavior is observed as predicted by Equation (4.66).

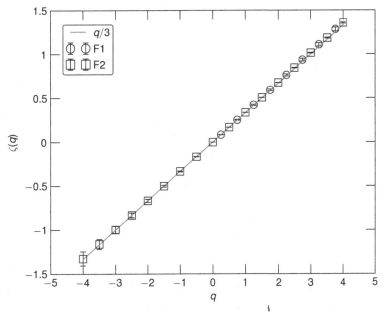

Figure 4.19 Experimental scaling exponent $\zeta(q)$ provided by formula (4.65) (denoted as F1, ○) and (4.67) (denoted as F2, □). The theoretical value $q/3$ is illustrated by a solid line.

is observed for all q considered here, verifying the scaling prediction (4.66). The corresponding scaling exponent $\zeta(q)$ curve is then illustrated in Figure 4.19, in which the theoretical prediction $q/3$ is represented by a solid line. Note that definitions (4.65) (denoted as F1) and (4.67) (denoted as F2) have the same performance except that the latter one can retrieve the negative moments, $q < 0$ (Huang, 2014).

4.6 An interpretation in a time-frequency analysis frame

To understand better the filter property of the detrending procedure and DSFs, we shall introduce here a weight function $W(\tau,f)$:

$$M_2(\tau) = \int_{-\infty}^{+\infty} W(\tau,f)E(f)df \qquad (4.68)$$

in which $E(f)$ is the Fourier power spectrum of $x(t)$, and $M_2(\tau)$ is a second-order moment, which could be one of $\mathcal{D}_2(\tau)$ or $\mathcal{B}_2(\tau)$, or $S_2(\tau)$. The weight function $W(\tau,f)$ characterizes the contribution of the Fourier component to the corresponding second-order moment. Note that an integral constant is neglected in the Equation (4.68). For the second-order SFs, we have the following weight function (Frisch, 1995; Huang et al., 2010),

$$W_{SF}(\tau,f) = 1 - \cos(2\pi f \tau) \qquad (4.69)$$

For a scaling process, one usually has a fast decaying Fourier spectrum, i.e., $E(f) \sim f^{-\beta}$ with $\beta > 0$. Hence, the contribution from small-scales (resp. high frequency Fourier mode) decreases. For the detrended data, the corresponding weight function should ideally be the following:

$$W_{DA}(\tau,f) = \begin{cases} 0, & \text{when } f \leq f_\tau \\ 1, & \text{when } f > f_\tau \end{cases} \qquad (4.70)$$

The DSFs (resp. the combination of the DA and SF) have a weight function, i.e.,

$$W_{DSF}(\tau,f) = \begin{cases} 0, & \text{when } f \leq f_\tau \\ 1 - \cos(\pi f \tau), & \text{when } f > f_\tau \end{cases} \qquad (4.71)$$

Hence the large-scale fluctuation influence (resp. infrared effect) is removed or constrained. Figure 4.20 shows the corresponding $W(\tau, f)$ for the SF, detrending analysis, and DSF, respectively. The detrended scale τ is illustrated by a vertical line, i.e., $f_\tau = 1/\tau$. We note here that with the definition of Equation (4.64), $\mathcal{B}_2(\tau)$ provides a better compatible interpretation with the Fourier power spectrum $E(f)$ since we have $W(\tau,f_\tau) = 1$. This is the main reason why we define the increment with the half size of the window width τ.

Let us comment on Equation (4.68). The above argument is exactly valid for linear and stationary processes. In reality, the data are always nonlinear and nonstationary for several reasons; for further discussion, see Huang et al. (1998). Therefore, Equation (4.68) holds approximately for real data. Another comment should be emphasized here for the detrending procedure. Several approaches might be

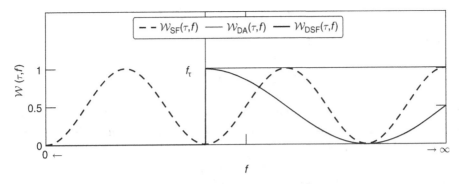

Figure 4.20 Illustration of the weight function $\mathcal{W}(\tau,f)$ for structure function analysis (dashed line), detrending analysis (thin solid line), and detrended structure-function analysis (thick solid line). The separation scale $f_\tau = 1/\tau$ is indicated by a vertical solid line.

applied to remove the trend (Wu et al., 2007; Bashan et al., 2008). However, the trend might be linear or nonlinear. Therefore, different approaches might provide different performances for different type of trends. In this study, we are only considering the zeroth- or first-order polynomial detrending procedure, which is found to be efficient for many types of data.

The above argument also implies that with a proper design of the weight function $\mathcal{W}(\tau,f)$, not only the large-scale effect (resp. infrared effect) but also the small-scale effect (resp. ultraviolet effect) can be removed or constrained to provide a better performance for extracting the scaling behavior of a given data set.

4.7 Wavelet-based methodologies

To overcome the global limitation of the Fourier transform, several methods have been proposed. These include: short-time Fourier transform (also known as spectrogram); proper orthogonal decomposition (also known as principal component analysis, or Karhunen-Lòeve transform, or empirical orthogonal function, etc.,); Wigner-Ville distribution (also known as Heisenberg wavelet); wavelet transform, to mention a few. In general, all these methodologies have been proposed to handle nonstationary and nonlinear data, and therefore succeed in some aspects but fail in other aspects. For example, none of these methods can handle nonlinearities of the given process, which cause the high-order harmonic problem (Cohen, 1995; Flandrin, 1998; Huang et al., 1998; see also a discussion in Section 1.3.2). This is partially because these methods are deeply related with the classical Fourier analysis, in which the basis function is proposed *a priori* (Huang et al., 1998). Concerning the scaling analysis, we shall discuss below only the wavelet transform.

4.7.1 Wavelet transform

The wavelet transform is defined as:

$$W(a,t) = \frac{1}{\sqrt{a}} \int_{-\infty}^{+\infty} x(t') \psi\left(\frac{t'-t}{a}\right) dt' \qquad (4.72)$$

in which ψ is the so-called mother wavelet, $a > 0$ is the dilation of the mother wavelet (resp. scale) and W is the wavelet coefficient (Daubechies, 1992). An individual wavelet is defined as:

$$\psi(a,t,t') = \frac{1}{\sqrt{a}} \psi\left(\frac{t'-t}{a}\right) \qquad (4.73)$$

To be a wavelet, the function ψ has to satisfy several mathematical conditions, such as the admissibility condition, a compact support range in both physical domain and spectral domains, etc. (Mallat, 1999; Daubechies, 1992; Meyer, 1995). The compact support property thus endows a capability to isolate nonstationary events. We shall not go further in this book on the details of the wavelet theory. For more information, we refer the readers to several classical books by Mallat (1999), Daubechies (1992), and Meyer (1995).

Figure 4.21 demonstrates the idea of wavelet transform with different scales. It presents a multiscale decomposition of a given data series. More precisely, for a wavelet with a given scale a, it is then scanned over all the time domain to retrieve the time-dependent wavelet coefficient $W(a,t)$. Note that the scale a can be associated with a characteristic frequency f due to the compact support property in spectral domain (Cohen, 1995; Daubechies, 1992). Figure 4.22 shows a shape of the Meyer wavelet and the corresponding Fourier power spectrum, illustrating the idea of the compact support not only in the physical domain, but also in the spectral domain. The compact support property in the spectral domain indicates a band-pass property of the wavelet.

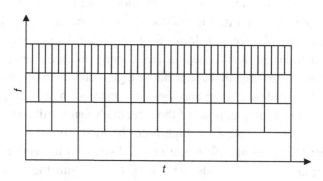

Figure 4.21 Illustration of a wavelet transform with different scales. It presents a multiscale decomposition of a given data series.

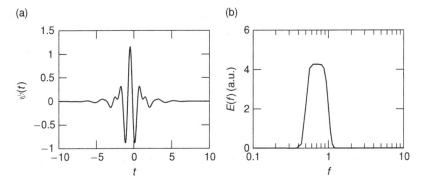

Figure 4.22 a) Illustration of a Meyer wavelet. b) The measured Fourier power spectrum. It shows a compact support property not only in the physical domain, but also in the spectral domain.

Since their introduction in the 1980s, wavelets have been widely used in data analysis and turbulence research (Muzy et al., 1991; Mallat and Hwang, 1992; Farge, 1992; Muzy et al., 1993; Farge et al., 1996; Arrault et al., 1997; Mallat, 1999; Rodrigues Neto et al., 2001; Jaffard et al., 2007; Wendt et al., 2007; Lashermes et al., 2008; Serrano and Figliola, 2009). Several wavelet-based methods have been proposed by several researchers to retrieve the scaling exponents from a scaling time series. For example, wavelet coefficients (WC), wavelet transform modulus maxima (WTMM; Muzy et al., 1991; Mallat and Hwang, 1992; Muzy et al., 1993), wavelet leader (WL; Jaffard et al., 2007; Wendt et al., 2007; Lashermes et al., 2008), etc.

Note that the form of Equation (4.72) is the same as the one for Fourier analysis; see Equation (1.3). It means that the basis function ψ, namely the mother wavelet, has to be chosen before the data analysis. Or in other words, the wavelet is still an *a priori* method and shares the same spirit as the Fourier transform. It thus possesses inherently more or less the potential shortcomings of the Fourier analysis (Cohen, 1995; Huang et al., 1998).

4.7.2 Haar wavelet

Historically, the Haar sequence has been recognized as the first wavelet basis (Haar, 1910). It is also recognized as the simplest wavelet. The Haar sequence is defined as:

$$\psi(t) = \begin{cases} 1 & 0 \leq t < 1/2, \\ -1 & 1/2 \leq t < 1 \\ 0 & \text{otherwise} \end{cases} \quad (4.74)$$

The Haar wavelet is simple to perform. It can be linked with the increment operator $\Delta_\ell x(t) = x(t+\ell) - x(t)$, in which ℓ is the separation scale and is $\ell = a/2$. Suppose we have a time series $[x_1, x_2, \cdots x_n]$. The Haar wavelet transform with a scale a at center x_k is then written as:

$$W(a,k) = \frac{1}{\sqrt{a}} \sum_{j=1}^{a/2} \left(x_{k-a/2+j} - x_{k+j} \right) \tag{4.75}$$

To compare with the structure function, see Equation (4.1); we redefine the Haar wavelet coefficient as:

$$W(a,k) = \frac{1}{a} \sum_{j=1}^{a/2} \left(x_{k-a/2+j} - x_{k+j} \right) = \langle \Delta_{a/2} x_i \rangle_{i \in [k-a/2, k+a/2]} \tag{4.76}$$

The above equation shows clearly that the Haar wavelet transform can be considered as a modification version of the structure function with a separation scale $\ell = a/2$ (Lovejoy and Schertzer, 2012). In other words, the classical structure function can be regarded as a poor man's wavelet. Haar wavelets are advocated as a new way to consider negative H values, hence more general than structure functions; see Lovejoy and Schertzer (2012, 2013).

Figure 4.23 shows the Haar wavelet, the simplest wavelet. By obtaining the Haar wavelet coefficient $W(\ell, k)$, one can define the qth-order moment as:

$$Z_q(\ell) = \langle |W(\ell, k)|^q \rangle \tag{4.77}$$

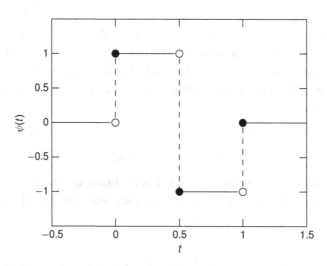

Figure 4.23 Illustration of the Haar wavelet; see Equation (4.74) for details. Note that the Haar wavelet can be linked with the classical structure function via the operation of the increment.

4.7 Wavelet-based methodologies

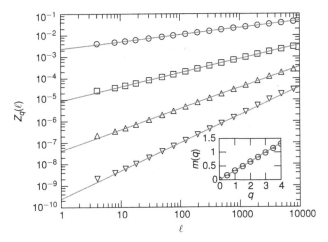

Figure 4.24 Measured qth-order moments $Z_q(\ell)$ for the fBm process with a Hurst number $H = 1/3$ computed using the Haar wavelet. Power-law behavior is observed. The inset shows the measured scaling exponent $m(q)$, in which the solid line indicates a theoretical prediction $q/3$.

in which for simplicity $W(a, k)$ is defined by Equation (4.76). For a scaling process, we expect a scaling behavior:

$$Z_q(\ell) \sim \ell^{m(q)} \tag{4.78}$$

in which $m(q)$ is the scaling exponent and comparable with the scaling exponent $\zeta(q)$ provided by the classical structure function. Figure 4.24 shows the measured $Z_q(\ell)$ for the fBm process with a Hurst number $H = 1/3$. Power-law behavior is observed. The corresponding measured scaling exponent $m(q)$ is shown as inset. It agrees with the theoretical value $q/3$ very well.

4.7.3 Wavelet transform modulus maxima

The idea of wavelet transform modulus maxima (WTMM) was first introduced by Mallat and Hwang (1992). It has been widely used in signal processing (Muzy et al., 1991, 1993). Let us consider here a discrete wavelet transform (DWT):

$$\psi(j, b) = \int_{\mathbb{R}} x(t) \varphi \left(2^{-j} t - b \right) dt \tag{4.79}$$

where φ is the chosen wavelet, $\psi(j, b)$ is the wavelet coefficient, b is the position index (resp. translation), j is the scale index, and $\ell = 2^j$ is the corresponding scale (Daubechies, 1992; Mallat, 1999). As we mentioned above the first way to detect the scale-invariant properties is to consider the wavelet coefficients

$$Z_q(j) = \langle |\psi(j,b)|^q \rangle \sim 2^{j\xi(q)} \tag{4.80}$$

where $\xi(q)$ are the corresponding scaling exponents.

Taking the wavelet coefficient $\psi(j,b)$, one can define a space-scale partition function for a given scale index j with the local maximum modulus $|\psi(j,b)| < |\psi(j,b_M)|$. We denote here $\psi(j,a) = |\psi(j,b_M)|$ for this maximum modulus. The qth-order moment is then defined as:

$$\mathcal{Z}_q(j) = \langle \psi(j,a)^q \rangle \tag{4.81}$$

where $q \in R$. For a scaling process, one has:

$$\mathcal{Z}_q(j) \sim 2^{j\zeta_W(q)} \tag{4.82}$$

where $\zeta_W(q)$ is the scaling exponent and is equivalent with the one provided by Equation (4.80). Note that one advantage of the WTMM is that it can retrieve the negative moments, $q < 0$.

4.7.4 Wavelet leaders

Every discrete wavelet coefficient $\psi(j,b)$ can be associated with the dyadic interval $\varrho(b,j)$

$$\varrho(j,b) = \left[2^j b, 2^j(b+1)\right] \tag{4.83}$$

Thus the wavelet coefficients can be represented as $\psi(\varrho) = \psi(j,b)$. Wavelet leaders are defined as

$$l(j,b) = \sup_{\varrho' \subset 3\varrho(j,b), j' \leq j} |\psi(\varrho')| \tag{4.84}$$

where $3\varrho(j,b) = \varrho(j,b-1) \cup \varrho(j,b) \cup \varrho(j,b+1)$ (Lashermes et al., 2005; Wendt et al., 2007). Thus power law behavior is expected

$$\mathbf{Z}_q(j) = \langle l(j,b)^q \rangle \sim 2^{j\xi(q)} \tag{4.85}$$

in which $\xi(q)$ is the corresponding scaling exponent. Its efficiency has been shown for various types of data sets (Jaffard et al., 2007; Wendt et al., 2007; Lashermes et al., 2008; Serrano and Figliola, 2009; Lashermes et al., 2005).

Figure 4.25 shows the measured qth-order moments for the WTMM $\mathcal{Z}_q(\ell)$ and WL $\mathbf{Z}_q(\ell)$ for the fBm process with a Hurst number $H = 1/3$. Power-law behavior is observed as expected. The corresponding scaling exponent $\zeta_W(q)$ and $\xi(q)$ are also shown. The theoretical value is illustrated by a solid line. Visually, the measured scaling exponent agrees well with the theoretical one.

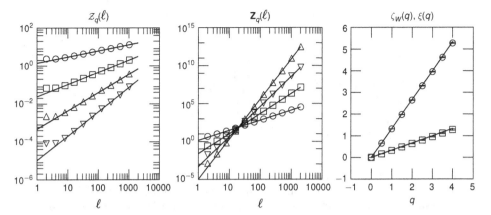

Figure 4.25 Measured qth-order moments for the fBm process with a Hurst number $H = 1/3$: a) Wavelet Transform Modulus Maximum $\mathcal{Z}_q(\ell)$. b) Wavelet leader $\mathbb{Z}_q(\ell)$; and c) the corresponding scaling exponents $\zeta_W(q)$ ($\square$) and $\xi(q)$ ($\bigcirc$). The theoretical value is illustrated by a solid line.

4.8 Empirical mode decomposition and Hilbert spectral analysis

A few years ago, a fully data-driven method, namely the Hilbert-Huang Transform (HHT) was proposed by Huang et al. (1998) to overcome the potential shortcoming of the Fourier-based analysis. Since it was introduced, this method has been widely used in many research fields and engineering applications, receiving several thousand citations in the literature. The so-called Hilbert-Huang Transform has two parts: empirical mode decomposition (EMD) and Hilbert spectral analysis (HSA). The first part is an empirical algorithm to decompose a given time series into a sum of series intrinsic mode functions (IMFs). The latter then associates each IMF mode to its instantaneous frequency using the classical Hilbert spectral analysis, in which the Hilbert transform is involved (Cohen, 1995). Below we present more details on this methodology.

4.8.1 Empirical mode decomposition

As mentioned above, in the real world most of the signals are multicomponent or multiscale, which means that different scales coexist simultaneously (Cohen, 1995; Huang et al., 1998, 1999). The multicomponent or multiscale property has been interpreted as fast oscillations superposed to slow ones at a local level (Rilling et al., 2003; Flandrin and Gonçalvès, 2004). For a giving decomposition method, a characteristic scale (CS) is defined implicitly or explicitly before the decomposition. For example, the CS in the classical Fourier analysis is a period λ of the sine wave, and in the wavelet transform it is the shape of the mother wavelet. In the HHT, it is then defined as the distance between two successive maxima (respectively

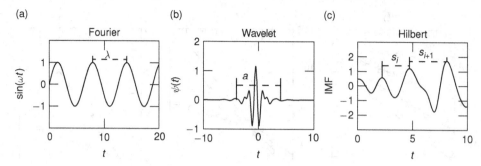

Figure 4.26 Illustration of the characteristic scale for different methodologies: a) Fourier transform; b) wavelet transform; and c) Hilbert-Huang transform. Note that only the Hilbert-based method allows both frequency- and amplitude-modulation.

minima) points (see Figure 4.26). Note that for the Fourier analysis, there is the assumption of the existence of a component existing on the whole data span. There is therefore no frequency modulation or amplitude modulation. Conversely, the wavelet transform allows amplitude modulation, since the mother wavelet possesses a compact support range in both physical and spectral domains. The HHT allows both frequency and amplitude modulations by the definition of its CS.

In HHT, Intrinsic Mode Function (IMF) is proposed to represent each mono-component signal that satisfies the definition of the CS. A given time series is then considered as a sum of IMFs. To be an IMF, a time series has to satisfy the following two conditions: (i) the difference between the number of local extrema and the number of zero-crossings must be zero or one; (ii) the running mean value of the envelope defined by the local maxima and the envelope defined by the local minima is zero (Huang et al., 1998, 1999); see Figure 4.27 for a typical IMF.

A subpart of the Empirical Mode Decomposition (EMD) algorithm, called "sifting process," is designed to decompose a given signal into several IMF modes (Huang et al., 1998, 1999; Rilling et al., 2003). For a given time series $x(t)$, the first step of the sifting process is to extract all the local maxima (resp. minima) points. The upper envelope $e_{max}(t)$ and the lower envelope $e_{min}(t)$ are then constructed, respectively, for the local maxima and minima points by using a cubic spline algorithm or other algorithm (Huang et al., 1998; Rilling et al., 2003). The running mean between these two envelopes is defined as

$$m_1(t) = \frac{e_{max}(t) + e_{min}(t)}{2} \tag{4.86}$$

Thus the first component is estimated by

$$h_1(t) = x(t) - m_1(t) \tag{4.87}$$

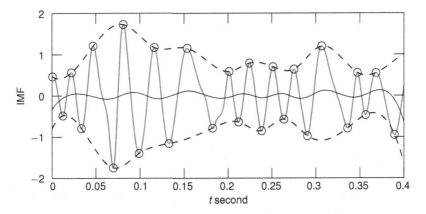

Figure 4.27 A typical Intrinsic Mode Function extracted from experimental data. The local extrema points are denoted by ○. The upper and lower envelopes are denoted by a dashed line.

Ideally, $h_1(t)$ should be an IMF as expected. In practice, $h_1(t)$ may not satisfy the above mentioned conditions. The function $h_1(t)$ is then taken as a new time series, and this sifting process is repeated j times, until $h_{1j}(t)$ is an IMF

$$h_{1j}(t) = h_{1(j-1)}(t) - m_{1j}(t) \tag{4.88}$$

The first IMF component $C_1(t)$ is then written as

$$C_1(t) = h_{1j}(t) \tag{4.89}$$

and the residual $r_1(t)$ as

$$r_1(t) = x(t) - C_1(t) \tag{4.90}$$

The sifting procedure is then repeated on the residual, until $r_n(t)$ becomes a monotonic function or at most has one local extreme point. This means that no more IMF can be extracted from $r_n(t)$. There are finally $n - 1$ IMF modes with one residual $r_n(t)$. The original signal $x(t)$ is rewritten at the end of the process as

$$x(t) = \sum_{i=1}^{n-1} C_i(t) + r_n(t) \tag{4.91}$$

To guarantee that the IMF modes retain enough physical sense, a certain stopping criterion has to be introduced to stop the sifting process properly. Different types of stopping criteria have been introduced by several authors (Huang et al., 1998, 1999; Rilling et al., 2003; Huang et al., 2003a; Huang, 2005). The first stopping criterion is a Cauchy-type convergence criterion. A standard deviation (SD), defined for two successive sifting processes, is written as

$$\text{SD} = \frac{\sum_{t=0}^{T} |h_{i(j-1)}(t) - h_j(t)|^2}{\sum_{t=0}^{T} h_{i(j-1)}^2(t)} \quad (4.92)$$

If a calculated SD is smaller than a given value, then the sifting stops, and gives an IMF. A typical value $0.2 \sim 0.3$ has been proposed based on Huang et al.'s experiences (Huang et al., 1998, 1999). Another widely used criterion is based on three thresholds α, θ_1, and θ_2, which are designed to guarantee globally small fluctuations, meanwhile taking into account locally large excursions (Rilling et al., 2003). The mode amplitude and evaluation function are given as

$$a(t) = \frac{e_{\max}(t) - e_{\min}(t)}{2} \quad (4.93a)$$

and

$$\sigma(t) = |m(t)/a(t)| \quad (4.93b)$$

so that the sifting is iterated until $\sigma(t) < \theta_1$ for some prescribed fraction $1-\alpha$ of the total duration, while $\sigma(t) < \theta_2$ for the remaining fraction. Typical values proposed by Rilling et al. (2003) are $\alpha \approx 0.05$, $\theta_1 \approx 0.05$ and $\theta_2 \approx 10\theta_1$, respectively based on their experience. A maximal iteration number (e.g., 300) is also chosen to avoid over-decomposing the time series.

The above described EMD algorithm performs the decomposition on a very local level in the physical domain without *a priori* basis. This means that the present decomposition is *a posteriori*: The basis is induced by the data itself (Huang et al., 1998, 1999; Flandrin and Gonçalvès, 2004). It is thus a scale-based decomposition. The EMD algorithm has been found to be a dyadic filter bank for various types of data (Flandrin and Gonçalvès, 2004; Wu and Huang, 2004; Huang et al., 2008a; Wu and Huang, 2010), showing the data-driven property as requested for the nonlinear and nonstationary data processing. Since its introduction, this method has attracted large interests in various research fields: waves (Schmitt et al., 2009; Hwang et al., 2003; Veltcheva and Soares, 2004); biological applications (Echeverria et al., 2001; Balocchi et al., 2004; Ponomarenko et al., 2005); financial studies (Huang et al., 2003b; Li and Huang, 2014); meteorology and climate studies (Huang et al., 2009a; Coughlin and Tung, 2004; Jánosi and Müller, 2005; Molla et al., 2006; Solé et al., 2007; Wu et al., 2007); mechanical engineering (Loh et al., 2001; Chen et al., 2004); acoustics (Loutridis, 2005); aquatic environment (Schmitt et al., 2007), and turbulence (Huang et al., 2008a, 2013; Tan et al., 2014; Huang et al., 2014), to mention a few. More detail about the EMD algorithm can be found in several methodological papers (Huang et al., 1998, 1999; Rilling et al., 2003; Flandrin and Gonçalvès, 2004; Flandrin et al., 2004; Huang, 2005).

4.8.2 Hilbert spectral analysis

With extracted IMF modes, one can apply the associated Hilbert spectral analysis to each component C_i, in order to extract the energy time-frequency information from the data (Huang et al., 1998, 1999; Long et al., 1995). The Hilbert transform is a singularity transform, which is written for an IMF mode $C(t)$ as

$$\tilde{C}(t) = \frac{1}{\pi} P \int \frac{C(t')}{t - t'} dt' \qquad (4.94)$$

where P means the Cauchy principle value (Cohen, 1995; Huang et al., 1998; Long et al., 1995). For each IMF mode function $C_i(t)$, one can then construct the analytical signal (Cohen, 1995), $\mathbb{C}_i(t)$, as

$$\mathbb{C}_i(t) = C_i(t) + j\tilde{C}_i(t) = \mathcal{A}_i(t) e^{j\theta_i(t)} \qquad (4.95)$$

where

$$\begin{cases} \mathcal{A}_i(t) = [C_i(t)^2 + \tilde{C}_i^2(t)]^{1/2} & \text{amplitude function} \\ \theta_i(t) = \arctan\left(\frac{\tilde{C}_i(t)}{C_i(t)}\right) & \text{phase function} \end{cases} \qquad (4.96)$$

Hence the instantaneous frequency is defined as the derivative of the phase function,

$$\omega_i(t) = \frac{1}{2\pi} \frac{d\theta_i(t)}{dt} \qquad (4.97)$$

The original signal is finally represented (excluding the residual $r_n(t)$) as

$$x(t) = \mathrm{R} \sum_{i=1}^{N} \mathcal{A}_i(t) e^{j\theta_i(t)} = \mathrm{R} \sum_{i=1}^{N} \mathcal{A}_i(t) e^{j \int \omega_i(t) dt} \qquad (4.98)$$

where "R" means real part. The above procedure is the classical Hilbert spectral analysis (Cohen, 1995; Flandrin, 1998). The combination of EMD and HSA is also called Hilbert-Huang Transform (HHT) by some authors (Huang, 2005). In general, this Hilbert-based transform can be taken as a generalization of the Fourier transform, since it allows frequency modulation and amplitude modulation simultaneously (Huang et al., 1998, 1999).

The EMD-HSA methodology is a time-energy-frequency approach: it provides at each time t, n values of instantaneous amplitude $\mathcal{A}_i(t)$ and instantaneous frequency $\omega_i(t)$, where n is the number of modes. After the Hilbert spectral analysis, a time-frequency energy representation, $H(\omega, t) = \mathcal{A}^2(\omega, t)$ can be extracted, called the Hilbert spectrum. It represents a relation between time, frequency, and energy

(Long et al., 1995). The Hilbert marginal spectrum is defined as a marginal integration of $H(\omega, t)$

$$h(\omega) = \frac{1}{T} \int_0^T H(\omega, t) dt \qquad (4.99)$$

in which T is the time span for energy estimation. $h(\omega)$ is similar with the Fourier spectrum, and can be interpreted as the energy associated with each frequency. However, we have to underline here the fact that the definition of frequency ω is different from the definition in the Fourier frame (see Figure 4.26). For example, since HHT allows frequency modulation, $\omega(t)$ itself is a function of time. The instantaneous frequency has its own distribution, which may depend on the mechanism behind the data, and can be extracted experimentally (Huang, 2009; Huang et al., 2014). Thus the interpretation of the Hilbert marginal spectrum should be given caution (Huang et al., 1998, 1999).

4.8.3 Arbitrary-order Hilbert-based statistics

Arbitrary-order Hilbert spectrum

Equation (4.99) can be rewritten by considering a joint pdf $p(\omega, \mathcal{A})$ of the instantaneous frequency ω and the amplitude $\mathcal{A}$ for all of these IMF modes (Huang et al., 2008a,b; Huang, 2009). With the joint pdf $p(\omega, \mathcal{A})$, the Hilbert marginal spectrum (Equation [4.99]) is then rewritten as

$$h(\omega) = \int_0^{+\infty} p(\omega, \mathcal{A}) \mathcal{A}^2 d\mathcal{A} \qquad (4.100)$$

Note that the above definition is no more than a second-order statistical moment. It is thus natural to generalize this approach to arbitrary-order moment $q \geq 0$ (Huang et al., 2008a,b; Huang, 2009)

$$\mathcal{L}_q(\omega) = \int_0^{+\infty} p(\omega, \mathcal{A}) \mathcal{A}^q d\mathcal{A} \qquad (4.101)$$

In the case of scale invariance, the following power-law behaviour is expected:

$$\mathcal{L}_q(\omega) \sim \omega^{-\xi(q)} \qquad (4.102)$$

in which $\xi(q)$ is the Hilbert-based scaling exponent function. Due to the integration operator, $\xi(q) - 1$ can be associated with $\zeta(q)$ from structure function analysis (Huang et al., 2008a; Huang, 2009).

We provide here some comments on Equation (4.101). There are at least two special values of q. One is $q = 0$. The corresponding zero-order Hilbert marginal

4.8 Empirical mode decomposition and Hilbert spectral analysis

spectrum is $\mathcal{L}_0(\omega) = p(\omega) = \int p(\omega, \mathcal{A}) d\mathcal{A}$, namely the marginal pdf of the instantaneous frequency ω. As mentioned above the distribution $p(\omega)$ can be different for different processes. In practice, for various experimental data, an inverse power-law behavior was found, i.e., $p(\omega) \sim \omega^{-1}$ (Huang et al., 2008a, 2009a, 2010). The second special case is $q = 2$, corresponding to the second-order Hilbert spectrum. As discussed above, it represents the energy associated with each frequency.

ω-conditioned statistics

The statistics can also be defined another way, based on ω. Note that each IMF mode $C_i(t)$ can be associated with its instantaneous frequency $\omega_i(t)$. A ω-conditioned statistics is then defined as:

$$\mathcal{M}_q(\omega) = \langle |C_i(t)|^q \rangle_{t, \omega(t) = \omega} \tag{4.103}$$

For a scaling process, a power-law behavior is then expected:

$$\mathcal{M}_q(\omega) \sim \omega^{-\zeta_H(q)} \tag{4.104}$$

The scaling exponent $\zeta_H(q)$ can be compared with the one $\zeta(q)$ provided by the classical structure function. Note that other conditional statistics can be extracted. For example, one can associate the Lagrangian acceleration with different velocity structures using ω-conditioned statistics.

A main advantage of this Hilbert-based method is that it can constrain both the large-scale effect (infrared effect) and small-scale effect (ultraviolet effect), since this method can handle not only nonstationary events but also nonlinear events (Huang et al., 2010, 2013). We show this by an example: the Duffing equation as mentioned in Equation (1.5). Figure 4.28 shows the extracted IMF modes from the numerical solution of the Duffing equation. Three IMF modes with a residual are obtained. The first IMF mode $C_1(t)$ contains 99.6% of the energy with a mean frequency of 0.105 Hz, and represents the main variation of the Duffing equation. The second IMF mode $C_2(t)$ possesses a mean frequency of 0.04 Hz, corresponding to the external force. The third IMF mode $C_3(t)$ and the residual Resi might be boundary effects.

Figure 4.29 shows the instantaneous frequency provided by the Hilbert spectral analysis for the first two IMF modes. The dashed line illustrates the frequency of 0.105 Hz and 0.04 Hz. The first frequency corresponds to the mean frequency of the Duffing equation by using a peak-counting algorithm. The second one corresponds to the external force. Note that the former one is varying with time t. It possesses a complete oscillation within a period ~ 10. To emphasize this phenomenon, the instantaneous frequency $\omega_1(t)$ is then taken as a new time series.

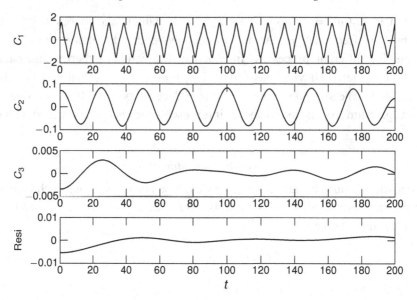

Figure 4.28 IMF modes from the nonlinear Duffing (Equation [1.5]). The first IMF mode with a mean frequency ~ 0.1 Hz captures the main variation of the Duffing equation. The IMF $C_2(t)$ corresponds to the external force. The IMF mode C_3 and the residual might be the boundary effect of the EMD algorithm.

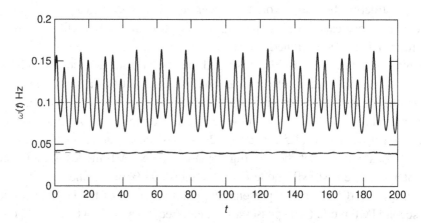

Figure 4.29 Measured instantaneous frequency $\omega(t)$ provided by the Hilbert spectral analysis for the first two IMF modes of the Duffing equation. The dashed line are for 0.105 Hz for the mean frequency of Duffing equation, and 0.04 Hz for the external force, respectively.

The HHT methodology is then applied to $\omega_1(t)$, and the instantaneous frequency $\omega(t)$ of the instantaneous frequency $\omega_1(t)$ is finally retrieved. Figure 4.30 shows the calculated instantaneous frequency $\omega(t)$ with a mean frequency $\langle \omega(t) \rangle \simeq 0.21$ Hz. This mean frequency is almost twice that of $\langle \omega_1(t) \rangle$. It confirms the existence of an

4.8 Empirical mode decomposition and Hilbert spectral analysis

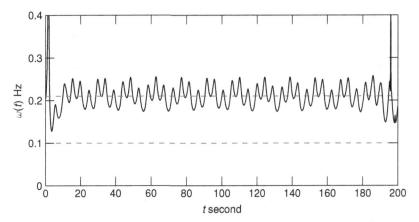

Figure 4.30 Measured instantaneous frequency $\omega(t)$ of the $\omega_1(t)$. The mean frequency is $\langle\omega(t)\rangle \simeq 0.21$ Hz. The dashed line indicates a mean frequency of $\langle\omega_1(t)\rangle \simeq 0.105$ Hz. The measured $\omega_1(t)$ possesses a complete oscillation within one period $\simeq 10$. This is the so-called intrawave frequency modulation, representing a nonlinear distortion.

oscillation within one period. The origin of this phenomenon has been recognized as a signature of nonlinear distortion (Huang et al., 1998, 1999). The classical Fourier analysis or Fourier-based analysis (e.g., wavelet transform) requires high-order harmonics to mimic this nonlinear distortion; these high-order harmonics have no physical meaning (see Figure 1.6 and discussion in Section 1.3.2). More examples of the nonlinear distortion can be found in Huang et al. (1998, 1999).

Let us now return to the question of nonstationary events. A Lagrangian trajectory with a vortex trapping event is considered here (see the typical event around $t/\tau_\eta = 100$ in Figure 1.5). Figure 4.31 shows the IMF modes $C_i(t)$ extracted from this Lagrangian trajectory. The strong nonstationary or intermittent event, namely vortex trapping, is then isolated by the EMD algorithm. The measured instantaneous frequency $\omega_1(t)$ of the first IMF mode with a mean frequency $\simeq 0.20$ (resp. around five times that of the Kolmogorov time scale) is shown in Figure 4.32, where, except for the vortex trapping event, the rest has been set as NaN. Visually, a frequency modulation is observed, indicating the nonlinear mechanism behind this time series.

A synthesized fBm process with a Hurst number $H = 1/3$ is then analyzed by using the above Hilbert-based method. Figure 4.33 shows the measured $\mathcal{L}_q(\omega)$ and $\mathcal{M}_q(\omega)$, respectively. Power-law behavior is observed for all q considered here. The measured scaling exponent $\xi(q)$ ($\bigcirc$) and $\zeta_H(q)$ ($\square$) are shown as an inset, in which the theoretical prediction is illustrated by a solid line. Visually, they agree well with the theoretical ones, showing the efficiency of the Hilbert-based methodology.

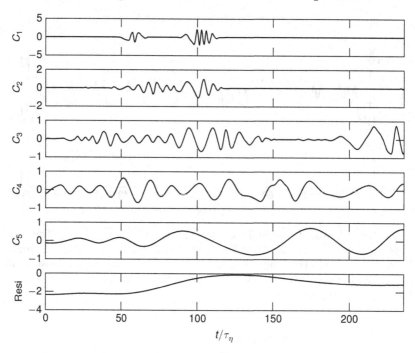

Figure 4.31 Extracted IMF modes $C_i(t)$ from a Lagrangian trajectory with a vortex trapping event. Visually, a vortex trapping event with a typical time scale $5\tau_\eta$ is isolated around $t/\tau_\eta = 100$.

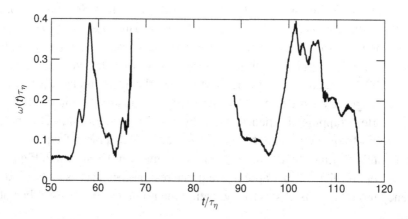

Figure 4.32 Extracted IMF modes $C_i(t)$ from a Lagrangian trajectory with a vortex trapping event. Visually, the vortex trapping event with a typical time scale $\sim 5\tau_\eta$ is isolated around $t/\tau_\eta = 100$. Note that the rest of the vortex trapping event has been set as NaN.

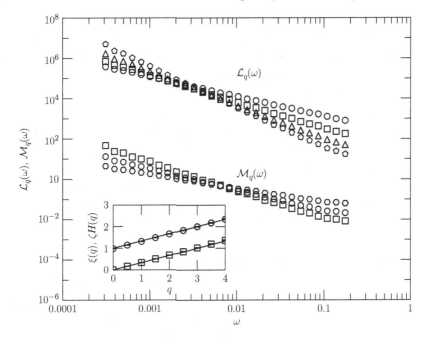

Figure 4.33 Experimental $\mathcal{L}_q(\omega)$ and $\mathcal{M}_q(\omega)$ for the fBm process with a Hurst number $H = 1/3$. Power-law behavior is observed for all q considered here. The inset shows the measured scaling exponent $\xi(q)$ (○) and $\zeta_H(q)$ (□), in which the solid line indicates the theoretical prediction.

4.9 General remarks on the scaling analysis methodologies

To summarize this chapter, several methodologies, especially those that have emerged in recent years, have been introduced. Their main properties have been discussed. For example, the classical structure function is suffering with both the infrared and ultraviolet effects, corresponding respectively to the influence of large-scale and small-scale motions. This causes from the mixture feature of the increment operator, which can be shown quantitatively by introducing a relative cumulative function (see Equation [4.8]). Several attempts to recover the shortcoming of the classical structure function have been made, including the autocorrelation function of increments, maximum pdf of increments, detrended structure function, etc.

As will be shown in the rest of this book, these approaches have been successful in several aspects, but they have their own disadvantages. For example, the autocorrelation function is no more than a second-order statistics, while the pdf scaling is comparable with the first-order structure function. Although the detrended structure function can constrain the large-scale influence, it still suffers from small-scale influences. Wavelet-based methodologies have been widely applied to different data sets; however, similarly with the Fourier analysis, they inherit the same potential shortcomings when dealing with nonlinear processes. For example, the discrete wavelet transform is not in agreement with two facts of the time-frequency representation of a time series. First, the scale of a time series belonging to a complex system, for example, a turbulent flow, is continuous in a statistical sense, but not discrete on several scales (Huang et al., 2008a; Huang, 2009). Second is that a certain scale may not exist all the time (Flandrin, 1998; Huang et al., 1998; Huang, 2009); see also the discussion in Chapter 5. Thus to represent a signal by using a discrete wavelet transform is not consistent with these physical aspects.

In practice, the Hilbert-based method can handle both nonlinear and nonstationary events, and constrain both the large-scale and small-scale influences. However, the first part of this method, namely empirical mode decomposition, still lacks a sound mathematical ground.

As shown in this chapter, for the simple synthesized fBm process, all methods provide almost the same performance. It is also true for synthesized multifractal processes (Huang et al., 2011a). The real data sets from the real world are different with synthesized ones, since different structures are hidden in the real data (Huang et al., 2014). Therefore, for different processes with different dynamics or intrinsic structures, different methodologies might have different performances. Our experience is expressed below. Before the data are analyzed, one has to check first the data set with the naked eye to see whether the large-scale fluctuations or energetic nonstationary events are visible or not. If yes, the classical structure function should be applied with caution. Quantitatively, the so-called relative cumulative function (Equation [4.8]) for the second-order structure function should be verified carefully to detect the influence of large-scale or small-scale motions. Another suggestion is to apply different methodologies to the same data to see whether they provide the same scaling exponents or not. If not, one has to check the data or method itself to see whether the conditions to apply the method are satisfied.

In the rest of the book, the above-mentioned methods are applied to several real data sets from different fields, to show their performances.

5
Applications: case studies in turbulence

In this chapter, the previous chapter's methods will be applied to several data sets obtained from the field of turbulence. The experimental velocity data was obtained from a wind tunnel experiment at the Johns Hopkins University, Baltimore, Maryland, USA. It includes also Lagrangian data obtained from a high-resolution direct numerical simulation (DNS), temperature field obtained from the Rayleigh-Bénard convection systems, and a vorticity field from a very high-resolution DNS.

5.1 Homogeneous turbulence

The experimental velocity data obtained from a wind tunnel experiment was performed by Prof. Meneveau's group with an active grid technique to achieve a high Reynolds number. The measurement locations were $x/M = 20$, $x/M = 30$, $x/M = 40$, and $x/M = 48$, where M was the mesh size and x was the downstream location. An X-type hotwire was used to measure the velocity, each time for a duration of 30 seconds (there were 30 realizations of time series measurements), with a sampling frequency of 40 kHz. In this book, we shall only consider the data obtained at the location $x/M = 20$. We recall here the main parameters of this data set. The Taylor's microscale based Reynolds number is $Re_\lambda \simeq 720$, the mean velocity is $12.0 \, \text{ms}^{-1}$, and the turbulent intensity is about 15.4%. We perform the analysis of this database in the temporal domain with the implicit use of the Taylor's frozen hypothesis to convert the temporal data to the spatial one. More detail of this database can be found in Kang et al. (2003).

5.1.1 Fourier power spectrum and second-order structure function

Figure 5.1 shows the experimental Fourier power spectrum $E(f)$ for both the longitudinal (solid line) and transverse (dashed line) velocity components. The Fourier power spectrum $E(f)$ is estimated with a realization of $2^{15} = 32768$

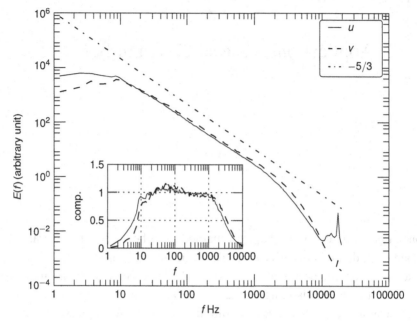

Figure 5.1 Experimental Fourier power spectrum $E(f)$ for the longitudinal (solid line) and the transverse (dashed line) velocities, in which the Kolmogorov's $-5/3$-law is shown as a dash-dotted line. Power-law behavior is observed on nearly a two-decade range $10 < f < 1000$ Hz with the scaling exponents $\beta = 1.63 \pm 0.01$ and 1.61 ± 0.01, respectively, for the longitudinal and transverse velocity components. The inset shows the compensated curves by using the fitted scaling exponents to emphasize the observed power-law behavior.

data points each without overlap. Power-law behavior is visible for both curves on nearly two decades of frequency range $10 < f < 1000$ Hz with the scaling exponents $\beta = 1.63 \pm 0.01$ for the longitudinal velocity, and $\beta = 1.61 \pm 0.01$ for the transverse velocity. Note that these two values are slightly smaller than the Kolmogorov value of 5/3, which could be considered as a finite Reynolds number effect. Power-law behavior is thus expected for at least the second-order structure functions on the time scales $0.001 < \tau < 0.1$ second; see the mathematical relation in Equation (4.5). A measurement noise is visible for the longitudinal velocity around $f_N \simeq 20{,}000$ Hz. In the energy-contain range, for example, $1 < f < 10$ Hz, the measured $E(f)$ is different for the longitudinal and transverse velocity components. More precisely, the longitudinal component contains more energy than the transverse one. As shown in this section, the scaling exponents could be biased due to the large-scale contamination, providing a different scaling exponent for these two velocity components.

Figure 5.2 shows the measured second-order SFs for both the longitudinal (○) and the transverse (□) velocity components. Power-law behavior is observed on

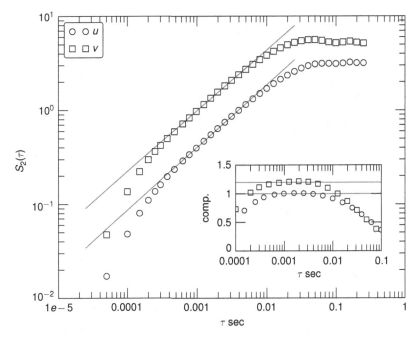

Figure 5.2 Experimental second-order structure functions $S_2(\tau)$ for the longitudinal ($\bigcirc$) and transverse ($\square$) velocity components. Power-law behavior is observed on the range $0.0005 < \tau < 0.005$ second, respectively, with scaling exponents $\zeta(2) = 0.67 \pm 0.01$ and $\zeta(2) = 0.65 \pm 0.02$. The power-law range corresponds to a frequency range of $200 < f < 2000$ Hz. The inset shows the compensated curves to emphasize the observed power-law behavior. For display clarity, these curves have been vertical shifted.

the range $0.0005 < \tau < 0.005$ second, corresponding to a frequency range $200 < f < 2000$ Hz. Note that this inertial range is shorter than the one predicted by the Fourier analysis due to the scale mixture problem of SFs as mentioned in Chapter 4.

We now turn to the relative contribution function $\mathcal{P}(\tau,f)$, namely the cumulative function, obtained from the real experimental data. The cumulative function $\mathcal{P}(\tau,f)$ is written as:

$$\mathcal{P}(\tau,f) = \frac{\int_0^f E(f')\left(1 - \cos(2\pi f'\tau)\right) df'}{\int_0^{+\infty} E(f)\left(1 - \cos(2\pi f\tau)\right) df} \times 100\% \tag{5.1}$$

As mentioned above, this characterizes a relative contribution to the second-order structure-function $S_2(\tau)$ from the frequency band $[0,f]$ (Huang et al., 2010). The Fourier power spectrum $E(f)$ is provided by the experimental data, shown in Figure 5.1. Figure 5.3 shows the measured cumulative function $\mathcal{P}(\tau,f)$ for various separation scales in the inertial range predicted by the Fourier power spectrum: $\tau = 0.005$

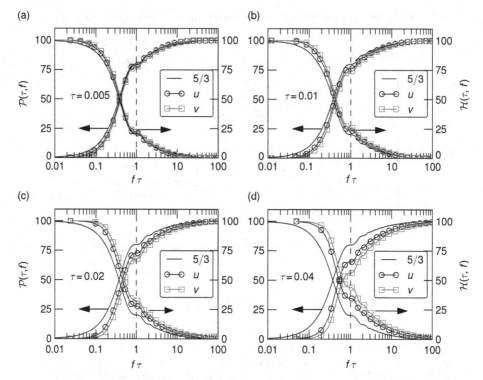

Figure 5.3 Experimental cumulative function $\mathcal{P}(\tau,f)$ for various separation time scales in the inertial range: a) $\tau = 0.005$ (resp. $f_o = 200$), b) $\tau = 0.01$ (resp. $f_o = 100$), c) $\tau = 0.02$ (resp. $f_o = 50$), and d) $\tau = 0.04$ (resp. $f_o = 25$). The frequency has been rescaled as $f\tau$. For comparison, $\mathcal{P}(\tau,f)$ for a Kolmogorov power-law is illustrated as a solid line.

($f_o = 200$), $\tau = 0.01$ ($f_o = 100$), $\tau = 0.02$ ($f_o = 50$), and $\tau = 0.04$ ($f_o = 25$), respectively. For comparison, the curve obtained from the pure Kolmogorov spectrum, i.e., $E(f) \sim f^{-5/3}$, is illustrated by a solid line. Visually, with the increase of the separation scale, the measured $\mathcal{P}(\tau,f)$ deviates more from the theoretical curve, indicating a more serious mixture of scales. To quantify the contribution from the small-scale (the high frequency part), we introduce here a co-cumulative function,

$$\mathcal{H}(\tau,f) = 100 - \mathcal{P}(\tau,f) \qquad (5.2)$$

This characterizes a relative contribution from the frequency band $[f, +\infty]$, emphasizing the contribution from small scales. The experimental $\mathcal{H}(\tau,f)$ is also shown in Figure 5.3. It behaves similarly to the measured $\mathcal{P}(\tau,f)$, hence, both measured $\mathcal{P}(\tau,f)$ and $\mathcal{H}(\tau,f)$ imply a serious contamination problem of large-scale structures. One consequence of the scale mixture problem is that the SFs provide a shorter inertial range than the one predicted by the Fourier transform. We also

5.1.2 High-order structure functions

We now turn to the high-order structure functions. Figures 5.4 a and c show the measured $S_q(\tau)$, and b) and d) the corresponding compensated curve using the fitted scaling exponents $\zeta(q)$, for example, $S_q(\tau)\tau^{-\zeta(q)}$. Power-law behavior is observed for all qs considered here, i.e., $0 \le q \le 6$. The compensated curves confirm the observed scaling behavior on the range $0.0005 < \tau < 0.005$ seconds, correspond-

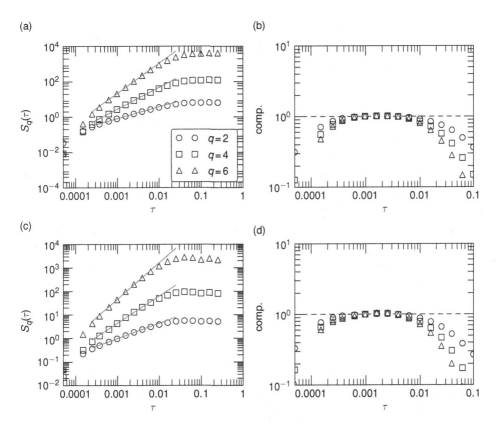

Figure 5.4 Experimental structure function $S_q(\ell)$: a) $S_q(\ell)$ for the longitudinal velocity; b) compensated curves using the fitted scaling exponent; c) $S_q(\ell)$ for the transverse velocity; and d) the corresponding compensated curves. A nearly one decade power-law behavior is observed on the range $0.0005 < \tau < 0.005$ for both velocity components.

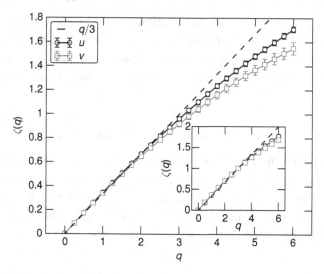

Figure 5.5 Measured scaling exponents $\zeta(q)$ of the longitudinal (○) and transverse (□) velocity components on the power-law range $0.0005 < \tau < 0.005$ for $0 \le q \le 6$. The inset shows the extended-self-similarity scaling exponents. The Kolmogorov nonintermittent value $q/3$ is indicated by a dashed line.

ing to a frequency range $200 < f < 2000$ Hz. One can find that the length of the inertial range is decreasing with q. We thus fix the inertial range as being $0.0005 < \tau < 0.005$, and extract the scaling exponent $\zeta(q)$ by using a least-square fitting algorithm. Figure 5.5 displays the measured scaling exponent $\zeta(q)$ for the longitudinal (○) and transverse (□) velocity components, in which the Kolmogorov's 1941 prediction $q/3$ is illustrated by a dashed line (see Figure 5.6 for the singularity spectrum). A clear deviation from the nonintermittent case $q/3$ is evident. When $q \ge 2$, the scaling exponents of the transverse velocity are smaller than the ones of the longitudinal component, for example, $\zeta^v(q) < \zeta^u(q)$. It also holds for the extended-self-similarity scaling exponents. More precisely, if one applies the lognormal formula, for example, $\zeta(q) = q/3 - \mu/2(q^2/9 - q/3)$, one has the measured intermittency parameter $\mu^u = 0.23 \pm 0.01$ for the longitudinal velocity, and $\mu^v = 0.32 \pm 0.01$ for the transverse one. This phenomenon has been interpreted as coming from the fact that the transverse velocity possesses a more intermittent effect. Therefore, instead of the energy dissipation rate, one should take the enstrophy (square of the vorticity) in the refined similarity hypothesis for the transverse velocity component (Chen et al., 1997).

5.1.3 Autocorrelation function of velocity increments

We now turn to the autocorrelation function of velocity increments. We first check the location $\tau_o(\tau)$ of the minimum value of the autocorrelation function $\Gamma_\tau(\ell)$,

5.1 Homogeneous turbulence

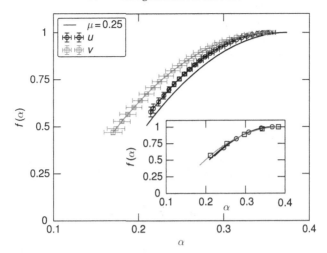

Figure 5.6 Measured singularity spectrum $f(\alpha)$ for the longitudinal ($\bigcirc$) and transverse ($\square$) velocity components. For comparison, the curve for the lognormal formula with an intermittency parameter $\mu = 0.25$ is illustrated by a solid line. The insets shows the measured $f(\alpha)$ provided by the extended-self-similarity method. For the same range of q, the transverse velocity has a wider α and $f(\alpha)$ than the longitudinal one. It has been interpreted as coming from the fact that the transverse velocity is more intermittent than the longitudinal one.

i.e., $\Gamma_\tau(\tau_o) = \min_\ell \{\Gamma_\tau(\ell)\}$, in which τ is the separation time scale, and ℓ is the time delay. As shown in Chapter 4, for a scaling process, the location τ_o is to be equal with the separation scale τ, i.e., $\tau_o(\tau) = \tau$; see details in section 4.2.2. Figure 5.7 shows the experimental $\tau_o(\ell)$ for the longitudinal ($\bigcirc$) and transverse ($\square$) velocity components, in which the inertial range $0.001 < \tau < 0.1$ sec predicted by the Fourier spectrum is shown by a vertical solid line. For both longitudinal and transverse velocity components, the relation $\tau_o(\tau) = \tau$ is valid not only on the inertial range, but also on the range far beyond the inertial range, for example, $\tau \gg 1$ sec, where the scaling behavior $E(f) \sim f^{-\beta}$ does not hold. Therefore, this relation is not purely a consequence of the scaling behavior.

Figure 5.8 shows the measured $\Gamma_o(\tau)$ for both longitudinal ($\bigcirc$) and transverse ($\square$) velocity components, in which the inset shows the compensated curve using the fitted exponents, and the solid line indicates a power-law fitting on the range $0.005 < \tau < 0.02$ sec with a scaling exponent 0.80 ± 0.02. To emphasize the observed power-law behavior, the inset shows the compensated curve using the fitted scaling exponent 0.80, in which the compensated curve for the second-order structure function is also shown by a solid line. It is found that the inertial range detected by the Γ_o is broader than the one provided by the structure function. This has been interpreted as coming from the fact that the autocorrelation function

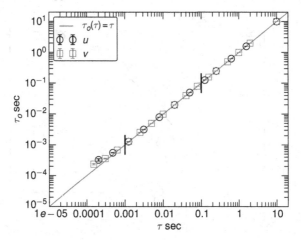

Figure 5.7 Measured location $\tau_o(\tau)$ for the minimum value of the autocorrelation function of velocity increments: the longitudinal (○) and transverse (□) velocity components, in which the inertial range $0.001 < \tau < 0.1$ second predicted by the Fourier spectrum is shown by a vertical solid line. The solid line indicates $\tau_o(\tau) = \tau$.

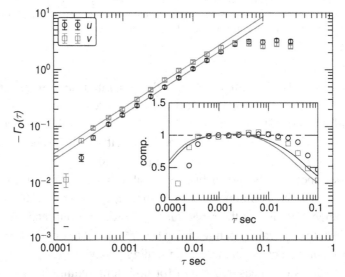

Figure 5.8 Measured $-\Gamma_o(\tau)$ for the longitudinal (○) and transverse (□) velocity components. The errorbar is a standard deviation from 120 realizations. Power-law behavior is observed on the range $0.0005 < \tau < 0.02$, with a scaling exponent 0.80 ± 0.02 for both velocity components. To emphasize the observed power-law behavior, the compensated curves using the fitted scaling exponent is shown as an inset. For comparison, the compensated curves for the second-order structure functions are also shown as thick (resp. longitudinal) and thin (resp. transverse) solid lines.

5.1 Homogeneous turbulence

of increments can constrain the large-scale influence (Huang et al., 2009b); see Fig. 4.10. However, the measured scaling exponent 0.80 is larger than the Kolmogorov value 2/3. A possible reason for this is that the autocorrelation function of increments is still influenced by the large-scale structures due to the nonlinear and nonstationary property of the turbulent flows.

5.1.4 Maximum pdf scaling of increments

Figure 5.9 shows the measured $p_{\max}(\tau) = \max\{p_\tau(x)\}$ for both longitudinal ($\bigcirc$) and transverse ($\square$) velocity components. The errorbar is obtained from 120 realizations. Power-law behavior, for example, $p_{\max}(\tau) \sim \tau^{-\alpha}$, is observed on the range $0.0005 < \tau < 0.01$ sec with a scaling exponent 0.36 ± 0.01. Note that the location of the maximum of $p_\tau(x)$ is located at $x \simeq 0$; see Equation (4.36). It indicates a

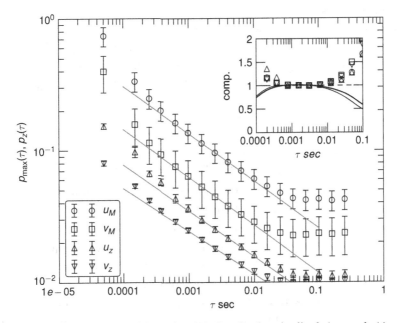

Figure 5.9 Measured $p_{\max}(\tau)$ and $p_z(\tau)$ for the longitudinal ($\bigcirc$, and $\triangle$) and transverse ($\square$, and $\triangledown$) velocity components. The errorbar is a standard deviation from 120 realizations. Power-law behavior is observed on the range $0.0005 < \tau < 0.01$ sec. The measured scaling exponents of $p_{\max}(\tau)$ are 0.36 ± 0.01 for both velocity components, and of $p_z(\tau)$ are $\alpha = 0.36 \pm 0.01$ and $\alpha = 0.33 \pm 0.01$ respectively for the longitudinal and transverse velocity components. To emphasize the observed power-law behavior, the compensated curves using the fitted scaling exponent is shown as inset. For comparison, the compensated curves with a scaling exponent $\zeta(1) = 0.35 \pm 0.01$ for the first-order structure functions are also shown as thick (resp. longitudinal) and thin (resp. transverse) solid lines.

potential scaling relation of the number of the zero-crossing points

$$p_z(\tau) \sim \tau^{-\alpha}, \quad p_z(\tau) = p_\tau(0) \tag{5.3}$$

In practice, we first calculate the number of zero-crossing points $N_z(\tau)$ for different separation scales τ. Then $p_z(\tau)$ is calculated as

$$p_z(\tau) = \frac{N_z(\tau)}{N(\tau)} \tag{5.4}$$

in which $N(\tau)$ is the total data length for the separation scale τ. The measured $p_z(\tau)$ is also shown in Figure 5.9 for both longitudinal ($\triangle$) and transverse ($\triangledown$) velocity components. Power-law behavior is evident as expected with a scaling exponent $\alpha = 0.36 \pm 0.01$ for the longitudinal velocity component, and $\alpha = 0.33 \pm 0.01$ for the transverse one. Note that the Kolmogorov 1941 value is 1/3 without intermittency correction. While for a high Reynolds number turbulent flow, the experimental value is around 0.37 (Benzi et al., 1995).

To emphasize the experimental power-law, the compensated curves using the fitted scaling exponent are shown as inset. A plateau is observed, confirming the existence of the power-law behavior. For comparison, the compensated curve for the first-order structure function is also shown as a solid line, using a scaling exponent 0.35 for both the velocity components. Visually, both pdf scaling and structure function provide the same scaling range, since both of them are based on the velocity increments.

5.1.5 Detrended structure functions

As discussed in section 4.5, the detrending process could remove or constrain the large-scale influence: the detrended kinetic energy $\mathcal{D}_2(\tau)$ obeys power-law behavior, for example, $\mathcal{D}_2(\tau) \sim \tau^{\beta-1}$, if the Fourier power spectrum has a power-law behavior, $E(f) \sim f^{-\beta}$; see Equation (4.63). Figure 5.10 shows the measured $\mathcal{D}_2(\tau)$ for both the longitudinal ($\bigcirc$) and transverse ($\square$) velocity components, in which the first-order detrending algorithm is applied. Power-law behavior is evident on the range $0.002 < \tau < 0.05$ seconds. The corresponding scaling exponents are 0.68 ± 0.02 and 0.65 ± 0.02 for the longitudinal and transverse cases respectively. Note that both scaling exponents are close to the Kolmogorov 1941 value 2/3. To emphasize the power-law behavior, the inset shows the compensated curve using the fitted scaling exponent.

We now turn to the detrended structure function. Figure 5.11 shows the measured $\mathcal{B}_q(\tau)$ for a) the longitudinal and b) transverse velocity components on the range $0 \le q \le 4$. Power-law behavior is observed on the range $0.002 < \tau < 0.02$ seconds. The scaling exponent $\zeta(q)$ is then estimated on this inertial range using a least-square fitting algorithm, which is indicated as a solid line in Figure 5.11.

5.1 Homogeneous turbulence

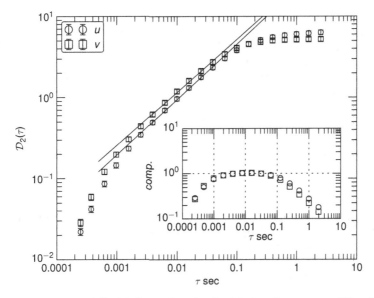

Figure 5.10 Measured $\mathcal{D}_2(\tau)$ for the longitudinal ($\circ$) and transverse ($\square$) velocity components. The errorbar is a standard deviation from 120 realizations. Power-law behavior is observed on the range $0.002 < \tau < 0.05$ sec with a measured scaling exponent 0.68 ± 0.02 and 0.65 ± 0.02 respectively for the longitudinal and transverse components. The inset shows the compensated curve using the fitted scaling exponent.

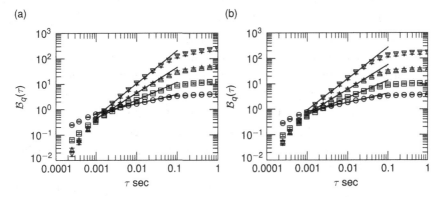

Figure 5.11 Measured detrended structure functions $\mathcal{B}_q(\tau)$ on the range $0 \leq q \leq 4$: a) the longitudinal velocity, and b) the transverse velocity. Power-law behavior is observed on the range $0.002 < \tau < 0.02$ sec.

Figure 5.12 a shows the experimental scaling exponents $\zeta(q)$, in which the Kolmogorov 1941 nonintermittent value $q/3$ is illustrated by a solid line. For comparison, a lognormal formula $q/3 - \mu/2(q^2/9 - q/3)$ with an intermittency parameter $\mu = 0.30$ (Huang, 2014) is shown by a dashed line. Graphically, the

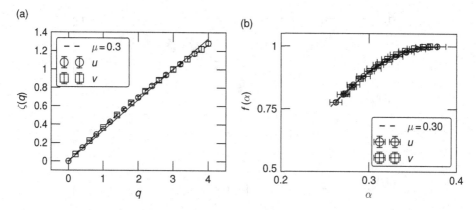

Figure 5.12 a) Experimental scaling exponents $\zeta(q)$ is extracted on the range $0.002 < \tau < 0.02$ seconds. b) The corresponding singularity spectrum $f(\alpha)$. For comparison, a lognormal formula with an intermittency parameter $\mu = 0.30$ is also shown as a dashed line.

experimental scaling exponents and the lognormal formula collapse very well, showing that there is no difference for the longitudinal and transverse velocity components, at least for the statistical order q up to 4. To emphasize this point, we calculate the singularity spectrum $f(\alpha)$, for example, $f(\alpha) = \min_q \{q\alpha - \zeta(q) + 1\}$, $\alpha = d\zeta(q)/dq$. Figure 5.12 b shows the measured $f(\alpha)$, which confirms that, statistically, there is no difference between the longitudinal and transverse velocity components. Furthermore, they can be described fully by the lognormal model with the intermittency parameter $\mu = 0.30$ (Huang, 2014).

Note that we obtain the scaling exponents here without applying the extended-self-similarity technique. While the traditional structure function analysis provides a different scaling exponent $\zeta(q)$ for the longitudinal and transverse velocity components (see Figure 5.6). It has been interpreted that the transverse velocity component is more intermittent than the longitudinal one. As a consequence, for the latter one, the energy dissipation rate involved the refined similarity hypothesis, while for the former one, the enstrophy is involved (Chen et al., 1997). Here we show that if the large-scale influence is removed/constrained, the scaling exponents $\zeta(q)$ and the corresponding singularity spectrum $f(\alpha)$ are then identical without statistical difference. Or, in other words, the small-scale isotropy might be contaminated by the large-scale anisotropy, which comes from either the geometric constrain in the laboratory experiment, or the large-scale forcing in the direct numerical simulation.

5.1.6 Detrended fluctuation analysis

Figure 5.13 shows the first-order detrended fluctuation analysis $\mathcal{F}_q(\tau)$ for both a) the longitudinal, and b) transverse velocity components on the range $0 \leq q \leq 6$.

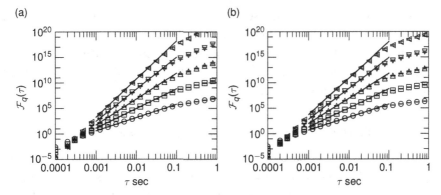

Figure 5.13 Experimental first-order detrended fluctuation functions: a) the longitudinal velocity, and b) the transverse velocity. Power-law behavior is observed on the range $0.002 < \tau < 0.02$ seconds.

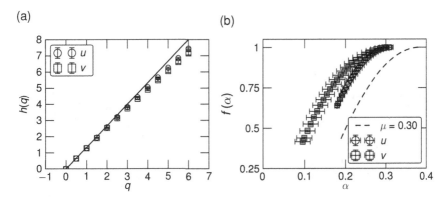

Figure 5.14 a) Experimental scaling exponents $\zeta(q)$ extracted on the range $0.002 < \tau < 0.02$ seconds, in which the solid line is the nonintermittent value $4q/3$. b) The corresponding singularity spectrum $f(\alpha)$. For comparison, a lognormal formula with an intermittency parameter $\mu = 0.30$ is also shown as a dashed line.

Power-law behavior is observed on the range $0.002 < \tau < 0.02$ seconds. This inertial range is the same as the one observed for the detrended structure function. However, it is shorter than the one predicted by the Fourier analysis (see Figure 5.1).

Figure 5.14 a shows the measured scaling exponent $h(q)$ for the longitudinal (○) and transverse (□) velocity components. For comparison, the Kolmogorov nonintermittent value $4q/3$ is shown by a solid line. Figure 5.14 b shows the measured singularity spectrum $f(\alpha)$, in which the lognormal model with an intermittency parameter $\mu = 0.30$ is also shown as a dashed line. Note that, similarly with the structure function, the measured singularity spectrum $f(\alpha)$ is different and indicates a more intermittent transverse velocity. As mentioned above, the detrended fluctuation analysis has been proposed to handle the nonstationary property of the data (see Section 4.4). However, in practice, the defined fluctuation function $\mathcal{F}_q(\tau)$ is still contaminated by the large-scale energetic structures, for example, forcing scale (Huang et al., 2011a). Or in other words, the scaling behavior measured by this method is still suffering the scale mixing problem.

5.1.7 Wavelet transform maximum modulus and wavelet leaders

Two widely used wavelet-based methods, wavelet transform modulus maximum and wavelet leaders, are considered here. Figure 5.15 shows the experimental qth-order wavelet-based moments for the longitudinal and transverse velocity components. Power-law behavior is observed on the range $0.0005 < \tau < 0.02$ seconds,

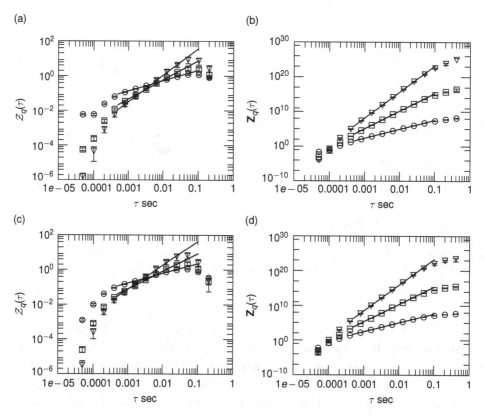

Figure 5.15 Experimental wavelet-based qth-order moments ($q = 2, 4, 6$): a) wavelet transform modulus maximum of the longitudinal velocity, b) wavelet leaders of the longitudinal velocity, c) wavelet transform modulus maximum of the transverse velocity, and d) wavelet leaders of the transverse velocity. Power-law behavior is observed on the range $0.0005 < \tau < 0.02$ seconds, corresponding to a frequency range $2000 < f < 50$ Hz.

5.1 Homogeneous turbulence

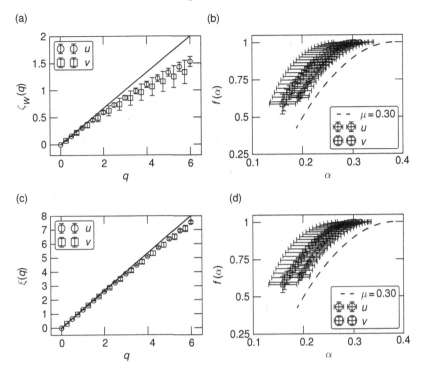

Figure 5.16 a) Experimental wavelet transform modulus maximum based scaling exponents $\zeta_W(q)$, in which the solid line is the nonintermittent value $q/3$, and b) the corresponding singularity spectrum $f(\alpha)$. c) Experimental wavelet leaders based scaling exponents $\xi(q)$, and d) the corresponding singularity spectrum $f(\alpha)$. For comparison, a lognormal formula with an intermittency parameter $\mu = 0.30$ is also shown as a dashed line.

corresponding to a frequency range $50 < f < 2000$ Hz. The wavelet-based scaling exponent is then estimated on this range. Figure 5.16 displays the measured scaling exponents and the corresponding singularity spectrum. For comparison, the lognormal formula with an intermittency parameter $\mu = 0.30$ is also shown as a dashed line.

Note that the wavelet transform modulus maximum and the wavelet leaders provide the same performance with an identical singularity spectrum. Similarly with the structure function analysis, detrended fluctuation analysis, are influenced by large-scale an isotropic structures. Therefore, they show a different scaling behavior or singularity spectrum for the longitudinal and transverse velocity components, and deviate from the lognormal formula. Once this large-scale anisotropy is constrained, the scaling exponent and the singularity spectrum become identical for the two velocity components (see Fig. 5.12 d).

5.1.8 Hilbert-Huang transform

We now apply the Hilbert-Huang Transform to this database. The decomposition is performed on each segment with 2^{14} data points without overlap. Below we present the main results.

Empirical mode decomposition

Figure 5.17 shows the number N of intrinsic mode functions retrieved for both the longitudinal (○) and transverse (□) velocity components. For comparison, a Gaussian distribution is shown as a dashed line. Experimentally, the velocity segment is decomposed into a sum of N intrinsic mode functions with a residual $r_N(t)$. The measured N is in the range $N \in [7, 12]$, which can be associated with a dyadic filter bank property of the empirical mode decomposition. More precisely, the measured N is usually limited as below,

$$N \leq \log_2(L) \tag{5.5}$$

in which L is the data length to be analyzed. Figure 5.18 shows the mean frequency of each mode in a semi-logy plot. The Fourier energy weighted mean frequency is defined as,

$$\bar{f}(n) = \frac{\int E_n(f) f df}{\int E_n(f) df} \tag{5.6}$$

in which $E_n(f)$ is the Fourier power spectrum of the nth retrieved mode. Graphically, an exponential law, for example, $\bar{f}(n) \sim \beta^{-n}$, is observed on the range

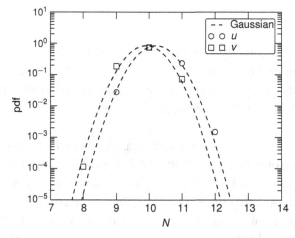

Figure 5.17 Experimental probability density function of the number of extracted intrinsic mode function of the longitudinal (○) and transverse (□) velocity components. For comparison, a Gaussian distribution is illustrated as a dashed line.

5.1 Homogeneous turbulence

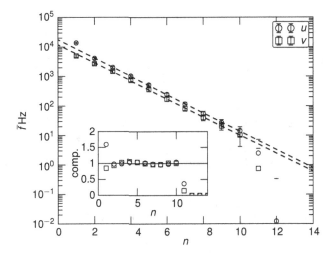

Figure 5.18 Experimental mean frequency $\bar{f}(n)$ of each intrinsic function mode. An exponential law, e.g., $\bar{f}(n) \sim \beta^{-n}$ is observed with a scaling exponent $\beta = 2.01 \pm 0.02$, indicating a dyadic filter bank of the empirical mode decomposition.

$2 \le n \le 10$ with a scaling exponent $\beta = 2.01 \pm 0.02$. It thus implies a dyadic filter bank property of this decomposition. More precisely, the measured mean frequency $\bar{f}(n)$ is half of the previous one. This dyadic-filter-bank-like behavior has been reported for various types of data, for example, white noise, turbulent velocity, daily river flow discharge, stock market index (Wu and Huang, 2004; Flandrin et al., 2004; Huang et al., 2008a, 2009a; Li and Huang, 2014), etc., to list a few. Note that the measured β could depend on the sifting number in the empirical mode decomposition algorithm (Wu and Huang, 2010). For example, once one increases the sifting number to infinity, the experimental β becomes closer to 1.

Joint probability density function

Figure 5.19 shows the experimental joint probability density function $p(\omega, \mathcal{A})$ for a) the longitudinal and b) transverse velocity components in a log-log plot. Graphically, there is a scaling trend in the inertial range $10 < \omega < 1000\,\mathrm{Hz}$. The scaling trend is further characterized by a skeleton, which is defined as,

$$p_{\max}(\omega) = p(\omega, \mathcal{A}_s(\omega)) = \max_{\mathcal{A}} \{p(\omega, \mathcal{A})|\omega\} \qquad (5.7)$$

The retrieved skeleton $\mathcal{A}_s(\omega)$ is illustrated as a $\bigcirc$.

Figure 5.20 shows the measured $\mathcal{A}_s(\omega)$. Power-law behavior is observed with a scaling exponent 0.33 ± 0.04. Note that this scaling exponent is close to the

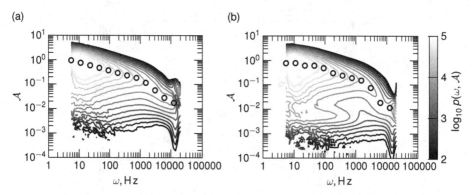

Figure 5.19 Experimental joint probability density function $p(\omega, \mathcal{A})$ for a) the longitudinal velocity component, and b) the transverse one. A scaling trend, namely skeleton, e.g., $p_{\max}(\omega) = p(\omega, \mathcal{A}_s(\omega)) = \max_{\mathcal{A}} \{p(\omega, \mathcal{A})|\omega\}$, $\mathcal{A}_s(\omega)$ is illustrated by a ○.

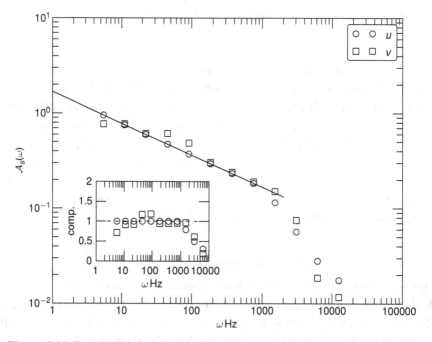

Figure 5.20 Experimental skeleton $\mathcal{A}_s(\omega)$. Power-law behavior is observed on the range $10 < \omega < 1000$ Hz with a Kolmogorov-like scaling exponent 0.33 ± 0.04. The inset shows the compensated curve to emphasize the observed scaling behavior.

5.1 Homogeneous turbulence

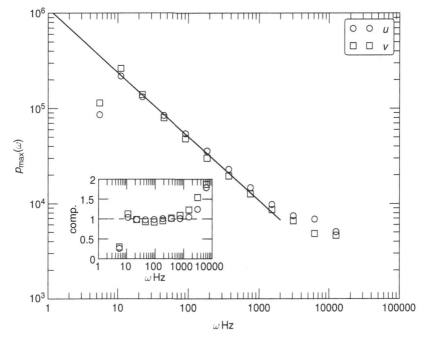

Figure 5.21 Experimental skeleton $p_{max}(\omega)$. Power-law behavior is observed on the range $10 < \omega < 1000\,\text{Hz}$ with a Kolmogorov-like scaling exponent 0.67 ± 0.04. The inset shows the compensated curve to emphasize the observed scaling behavior.

Kolmogorov-like value $1/3$. To emphasize this scaling behavior, the inset shows the compensated curve using the scaling exponent $1/3$. Figure 5.21 shows the corresponding $p_{max}(\omega)$. Power-law behavior is also shown on the inertial range $10 < \omega < 1000\,\text{Hz}$ with a scaling exponent 0.67 ± 0.04. Note that again this scaling value is close to the Kolmogorov value $2/3$. These skeleton scaling behaviors can be obtained using dimensional argument as shown in the next section.

Scaling of zeroth-order moment

Figure 5.22 shows the measured $p(\omega) = \int p(\omega, \mathcal{A})d\mathcal{A}$. Power-law behavior is observed on the range $10 < \omega < 3000\,\text{Hz}$ with a measured scaling exponent $\xi(0) = 0.98 \pm 0.01$. To emphasize the observed scaling behavior, the inset shows the compensated curve using the scaling exponent 1. The "-1"-like power-law has been reported for several different types of data, for example, fractional Brownian motion, surf-zone, financial index, etc. It seems that this scaling is a universal property of a stochastic process (Huang, 2009). Figure 5.23 reproduces the measured $\xi(q)$ from different processes, confirming a universal scaling of $\xi(0)$.

120   *Applications: case studies in turbulence*

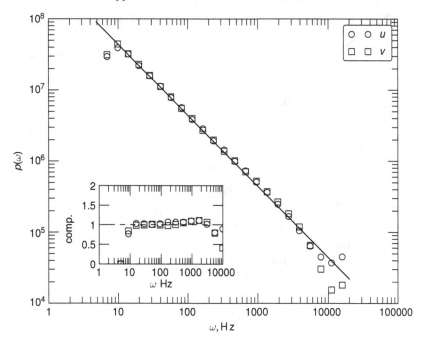

Figure 5.22 Experimental marginal distribution of $p(\omega)$. Power-law behavior is observed on the range $10 < \omega < 3000\,\text{Hz}$ with a scaling exponent $\xi(0) \simeq -1$. The inset shows the compensated curve to emphasize the observed scaling behavior using the scaling exponent -1.

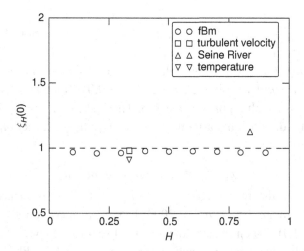

Figure 5.23 Experimental zeroth-order scaling exponent $\xi(0)$ obtained from different processes.

High-order moments

Figure 5.24 shows the measured qth-order moments $\mathcal{L}_q(\omega)$. Power-law behavior is observed on the range $10 < \omega < 1000\,\mathrm{Hz}$. The corresponding compensated curve shows a clear plateau, confirming the observed power-law behavior. The scaling exponent is then estimated on this range. Figure 5.25 a shows the measured scaling exponent $\xi(q) - 1$. Note that for discussion convenience, the zeroth-order scaling exponent $\xi(0)$ has been set as 1. Visually, the scaling exponents obtained from the longitudinal and transverse velocity components collapse, indicating the same intensity of the intermittency.

To emphasize this point, Figure 5.25 b shows the measured singularity spectrum $f(\alpha)$. Despite the fact that the errorbar of the transverse component is slightly larger than the longitudinal one, the measured $f(\alpha)$ is almost the same. For comparison, a curve obtained from a lognormal formula $\zeta(q) = q/3 - \mu/2(q^2/9 - q/3)$ with intermittency parameter $\mu = 0.30$ (solid line) and fitted value $\mu = 0.18$ are also shown. It is worth pointing out here again that the Hilbert-based method can

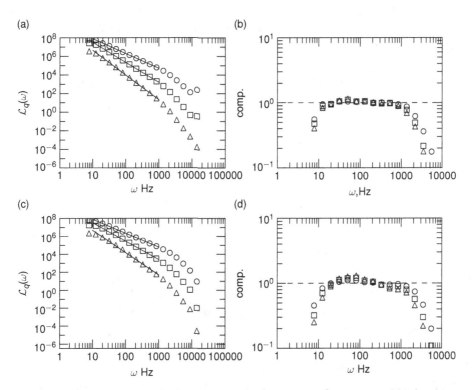

Figure 5.24 Experimental qth-order marginal moments $\mathcal{L}_q(\omega)$: a) and b) for the longitudinal velocity, c) and d) the transverse one. Power-law behavior is observed on the range $10 < \omega < 1000\,\mathrm{Hz}$.

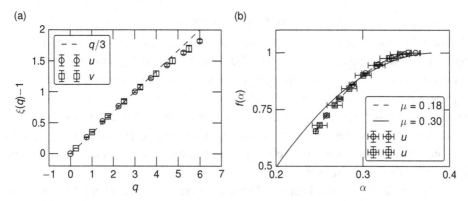

Figure 5.25 a) Experimental scaling exponents $\xi(q) - 1$; b) the measured singularity spectrum $f(\alpha)$. For comparison, a lognormal formula with intermittency parameter $\mu = 0.30$ (solid line) and a fitted one $\mu = 0.18$ (dashed line) are also shown.

separate different turbulent structures without *a priori* basis. The large-scale structure contamination, known as infrared effect is then constrained. The obtained scaling exponent is then the same for the longitudinal and transverse velocity components, implying the same intensity of the intermittency.

5.1.9 General remarks on scaling behaviors of the Eulerian velocity

As shown in Chapter 4, for a synthesized data, for example, fractional Brownian motion, all methods used above provide a comparable performance, since there is no intrinsic structures in the data (Huang et al., 2014). The intrinsic structure means a turbulent structure after another one; see discussion in the next subsection. This implies a distribution of scales in time or space.

For a real Eulerian velocity database, different methodologies show different performances. This is partially because of the scale mixture problem in the traditional approaches (Davidson and Pearson, 2005; Huang et al., 2010, 2011a, 2013). Therefore, the scaling behavior will be biased by the energetic structures, either large-scale ones, known as infrared effect, for example, forcing scale in the wind tunnel experiment, large-scale circulation in the Rayleigh-Bénard convection system, annual cycle in the collected geoscience data, etc., or small-scale ones, known as ultraviolet effect, such as vortex trapping events in the Lagrangian turbulence, etc. We also note that a pure scaling behavior is absent in the physical space. To overcome this problem, the extended-self-similarity is often applied, in which, instead of plotting the qth-order moment versus the scale τ, the qth-order moment is usually plotted versus the third-order moment (Benzi et al., 1993a, 1995). The power-law behavior is then clearer, and can be extracted more accurately. Despite the different performances of different methodologies, all of them confirm the intermittency or multifractal properties of the turbulent velocity. Moreover, once the influence of large-scale structures is constrained, the retrieved scaling exponents are almost the same for the longitudinal and transverse velocity components. Note that, previously the transverse velocity component was found to be more intermittent than the longitudinal one (Chen et al., 1997; Sreenivasan and Antonia, 1997). We show here that this conclusion may be revisited when different methods are applied.

5.2 Passive scalar

Now we consider here a scalar field θ advected by the turbulent velocity field. The governing equation is written as,

$$\partial_t \theta + \mathbf{u} \cdot \nabla \theta = \kappa \nabla^2 \theta \qquad (5.8)$$

where κ is the diffusivity, and $\mathbf{u}(\mathbf{x}, t)$ is the velocity field: the turbulent flow is not influenced by θ. The passive scalar field θ is then regarded as "passive scalar," known as passive scalar turbulence. The statistics of the scalar field θ is believed to be a **counterpart** of the velocity field (Warhaft, 2000; Shraiman and Siggia, 2000). For example, the generalization of the Kolmogorov's 1941 theory to the passive scalar results in a so-called Kolmogorov-Obukhov-Corrsin (KOC in short) scaling behavior (Obukhov, 1949; Corrsin, 1951),

$$S_q^\theta(\ell) = \langle \Delta \theta_\ell^q \rangle \sim \ell^{\zeta_\theta(q)} \qquad (5.9)$$

in which $\Delta \theta_\ell(x, t) = \theta(x + \ell, t) - \theta(x, t)$ is the temperature increment, ℓ is the separation scale, and $\zeta_\theta(q)$ is the inertial scaling exponent. The scaling exponent $\zeta_\theta(q)$ is expected to be the same as the one $\zeta(q)$ of the velocity field, possibly with a small modification since the passive scalar field θ is mainly advected by the turbulent flow.

Surprisingly, the measured scaling exponent $\zeta_\theta(q)$ is smaller than $\zeta(q)$ of the velocity field, indicating a more intermittent passive scalar field (Warhaft, 2000, see figure 11); see also Section 2.6. It has been interpreted as the effect of ramp-cliff structures, in which the temperature is decreasing gradually (ramp), and then is increasing rapidly (cliff; see Figure 5.26). The ramp is considered as a part of a large-scale structure, while the cliff is regarded as a part of a small-scale one. Due to the presence of the ramp-cliff structures, both the traditional Fourier analysis and the classical structure-function analysis are affected, leading to a more intermittent passive scalar field(Warhaft, 2000; Huang et al., 2010).

We consider here an experimental temperature data as passive scalar turbulence. The data are collected from a jet experiment performed by Prof. Gagne at Joseph

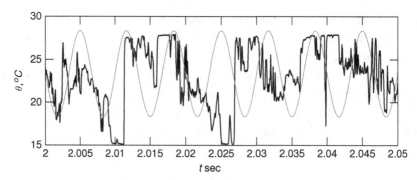

Figure 5.26 Illustration of a 0.05 second portion of temperature data $\theta(t)$ (thick solid line) as passive scalar turbulence. A typical ramp-cliff structure is observed with a gradually decreasing structure and a sharp increasing part. For comparison, a pure sine wave with 150 Hz is also shown. Visually, the ramp-cliff structure is significantly different with the sine wave.

Fourier University of Grenoble, France. The bulk Reynolds number is about $Re \simeq 60,000$. The Taylor's microscale Reynolds number is estimated as $Re_\lambda \simeq 250$. The initial temperatures of the two flows are $T_J = 27.8°C$ and $T = 14.8°C$. The measurement location is in the mixing layer and close to the nozzle of the jet. The sampling frequency is 50 kHz. The total data length is 10 sec, corresponding to 500,000 data points. Figure 5.26 shows a 0.05 sec portion temperature data, illustrating strong ramp-cliff structures. For comparison, a pure sine wave with a frequency 150 Hz is also shown. Obviously, the ramp-cliff structure is an energetic large-scale structure with a very sharp interface (Sreenivasan, 1991; Celani et al., 2000; Shraiman and Siggia, 2000; Warhaft, 2000). The profile of ramp-cliff structures deviates significantly from the sine wave. The Fourier-based methodologies thus require high-order harmonic components to represent this difference, in which the underlying idea is a linear asymptotic approximation (Cohen, 1995; Flandrin, 1998; Huang et al., 1998). The linear asymptotic approximation leads to an artificial energy flux from low frequencies (large-scale structures) to higher frequencies (small-scale structures), known as the infrared effect (Huang et al., 2010, 2011a, 2013). The measured scaling exponent β_θ of the Fourier spectrum $E_\theta(f)$ could thus be underestimated.

5.2.1 Fourier power spectrum and second-order structure function

Figure 5.27 a shows the measured Fourier power spectrum $E_\theta(f)$. Power-law behavior is observed on the range $100 < f < 1000$ Hz, corresponding to a time scale range $0.001 < \tau < 0.01$ seconds. The inset shows the compensated curve using

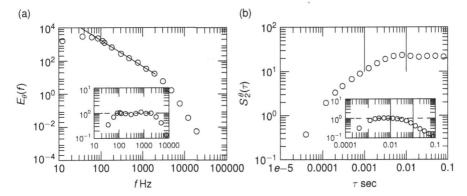

Figure 5.27 a) Experimental Fourier spectrum $E_\theta(f)$. Power-law behavior is observed on the range $100 < f < 1000$ Hz with a scaling exponent 1.56 ± 0.03. The inset shows the compensated curve to emphasize the scaling behavior. b) Experimental second-order structure function $S_2^\theta(\tau)$, in which the Fourier-based inertial range $0.001 < \tau < 0.01$ seconds is illustrated by a solid line. The inset shows the compensated curve using the KOC value $2/3$.

the fitted scaling exponent to emphasize the scaling behavior. The experimental Fourier scaling exponent is 1.56 ± 0.03. Note that this value is consistent with the one reported in Warhaft (2000). As discussed above, this scaling value could be underestimated due to the presence of ramp-cliff structures (Huang et al., 2010). Figure 5.27 b shows the corresponding second-order structure function $S_2^\theta(\tau)$, in which the Fourier-based inertial range $0.001 < \tau < 0.01$ seconds is illustrated by a solid line. Due to the influence of the ramp-cliff structures, there is a lack of power-law behavior. To show this more clearly, the inset shows a compensated curve using the KOC value $2/3$. There is no plateau observed. Note that once the influence of the energetic ramp-cliff structures is suppressed, one might retrieve the same scaling exponent as the one of the velocity (Huang et al., 2010, 2011a).

To quantify the contamination of the large-scale structures, the cumulative function $\mathcal{P}(\tau,f)$ defined in Equation (5.1) and the corresponding co-cumulative function $\mathcal{H}(\tau,f)$ are estimated for two separation time scales: $\tau = 0.002$ second and $\tau = 0.01$ second, corresponding to $f_o = 500$ Hz and 100 Hz in the inertial range. Note that the Fourier spectrum $E_\theta(f)$ is estimated with segments with a 2^{12} data points each without overlap and any window function. Figure 5.28 a shows the experimental $\mathcal{P}(\tau,f)$ and $\mathcal{H}(\tau,f)$. Visually, the second-order structure function $S_2^\theta(f)$ is strongly influenced by the large-scale structures, see $\mathcal{P}(\tau,f)$ at the point $f\tau = 1$. More precisely, $\mathcal{P}_1(\tau) = \mathcal{P}(\tau,f)|f\tau = 1$ characterizes the contribution from the scales $f\tau < 1$, namely large-scale contributions. Figure 5.28 b shows the measured $\mathcal{P}_1(\tau)$ and the corresponding $\mathcal{H}_1(\tau) = 100 - \mathcal{P}_1(\tau)$, which describes the contribution from small-scale structures, i.e., $f\tau > 1$. The Fourier-based inertial

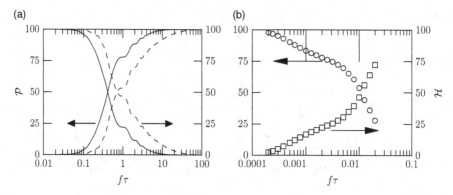

Figure 5.28 a) Measured cumulative function $\mathcal{P}(\tau,f)$ and the corresponding co-cumulative function $\mathcal{H}(\tau,f)$ at two time scales $\tau = 0.002$ sec and $\tau = 0.01$ sec, corresponding to $f_o = 500$ Hz and 100 Hz in the inertial range. b) The corresponding $\mathcal{P}_1(\tau)$ and $\mathcal{H}_1(\tau)$. The inertial range $0.001 < \tau < 0.01$ sec predicted by the Fourier power spectrum is illustrated by a solid line.

range is illustrated by a solid line. Note that in this inertial range, the contributions from large-scale parts are decreasing with τ, and are larger than 50%. The measured curve $\mathcal{P}_1(\tau)$ strongly depends on large-scale structures, especially energetic ones. The large-scale structures are known to be anisotropic, and depend on the flow conditions, such as the boundary conditions, forcing forms, etc. Therefore, $\mathcal{P}_1(\tau)$ and $\mathcal{H}_1(\tau)$ are expected to be different for different turbulent flows.

5.2.2 High-order structure function and extended-self-similarity

As shown above for the database considered, there lacks a power-law for the second-order structure function $S_2^\theta(\tau)$. This is also the case for higher order structure functions. Figure 5.29 a shows the measured structure functions $S_q^\theta(\tau)$. Due to the presence of the ramp-cliff structures, there is no power-law behavior for the measured $S_q^\theta(\tau)$. With the help of the extended-self-similarity, the structure-function scaling could be detected more accurately. Figure 5.29 b displays the ESS plot, i.e., $S_q^\theta(\tau)$ versus $S_3^\theta(\tau)$ on the range $0.001 < \tau < 0.01$ seconds, corresponding to the Fourier inertial range $100 < f < 1000$ Hz. Power-law is clearly visible. The scaling exponent $\zeta_\theta^S(q)$ is then extracted using a least-square fitting algorithm.

Figure 5.30a shows the measured extended-self-similarity scaling exponent $\zeta_\theta(q)$ (○), in which the scaling exponent $\zeta_\theta(q)$ compiled by Schmitt (2005) is also shown as □. For comparison with the velocity field, a lognormal formula with an intermittency parameter $\mu_\theta = 0.2$ is plotted as a solid line. Note that this intermittency parameter is widely used for the velocity field of high Reynolds number turbulent flows (Frisch, 1995). The current relative scaling exponent

5.2 Passive scalar

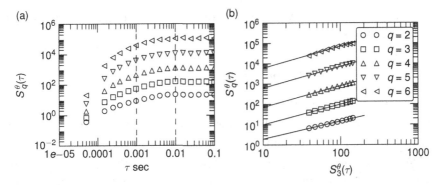

Figure 5.29 a) Measured high-order structure functions $S_q^\theta(\tau)$. The inertial range $0.001 < \tau < 0.01$ sec predicted by the Fourier power spectrum is illustrated by a solid line. b) The extended-self-similarity plot of the qth-order $S_q^\theta(\tau)$ versus $S_3^\theta(\tau)$ on the inertial range.

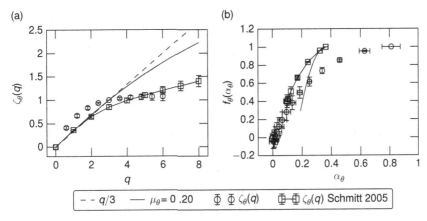

Figure 5.30 a) Measured extended-self-similarity scaling exponents $\zeta_\theta(q)$ (○). For comparison, the scaling exponent provided by a lognormal formula with an intermittency parameter $\mu_\theta = 0.20$ (solid line), and the one compiled by Schmitt (2005) (□) are also shown. b) The corresponding singularity spectrum $f_\theta(\alpha_\theta)$.

$\zeta_\theta(q)$ is approaching a constant, for example, $\zeta_\theta(\infty) = 1$, when $q \geq 3$, a saturation, which is believed to be an effect of the ramp-cliff structures (Celani et al., 2000). The compiled scaling exponent $\zeta_\theta(q)$ is smaller than the lognormal formula, implying a more intermittent passive scalar field. To emphasize this difference, the singularity spectrum $f_\theta(\alpha_\theta)$ is then calculated. Figure 5.30b displays the experimental singularity spectrum $f_\theta(\alpha_\theta)$ to emphasize the multifractality. The current singularity spectrum is wider than the compiled one and the one provided by the lognormal formula. It confirms again a more intermittent passive scalar

128 Applications: case studies in turbulence

field. However, as shown above, the structure-function is strongly influenced by the ramp-cliff structures.

5.2.3 Increment-based methods

Autocorrelation function of increments

Figure 5.31a shows the measured location $\tau_0(\tau)$ for the minimum value of autocorrelation function $\rho_\tau(\ell)$ of temperature increments, in which the inertial range predicted by the Fourier analysis is indicated by a solid line. Note that the relation $\tau_0(\tau) = \tau$ is valid on a large range of scales, even for the separation scale $\tau \gg 0.01$ sec. It seems again here that the power-law behavior is not a necessary condition for the Equation (4.23). A similar phenomenon has been observed for the turbulent velocity (see Figure 5.7).

Figure 5.31b shows the measured $\Gamma_\theta(\tau)$. Power-law behavior is clearly observed on the range $0.0002 < \tau < 0.005$ seconds, corresponding to a frequency range $200 < f < 5000$ Hz. The inset shows the compensated curve using a fitted scaling exponent. The measured scaling exponent is found to be 0.86 ± 0.02. This value is larger than the value $2/3$ for the second-order nonintermittent KOC one.

Maximum pdf of increments

Figure 5.32 shows the measured maximum pdf $p_{max}^\theta(\tau) = \max_x \{p_\tau^\theta(x)\}$ for different separation scales τ. A dual power-law behavior is observed on the range

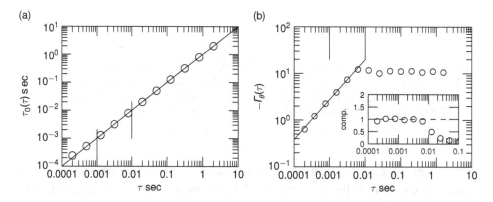

Figure 5.31 a) Measured location τ_0 for the minimum value of the autocorrelation function $\rho_\tau(\ell)$ of temperature increments. The solid line indicates $\tau_0(\tau) = \tau$. b) Experimental $\Gamma_\theta(\tau)$, in which the Fourier inertial range is indicated by a solid line. Power-law behavior is observed on the range $0.0002 < \tau < 0.005$ sec, corresponding to a frequency range $200 < f < 5000$ Hz with a scaling exponent 0.86 ± 0.02. Inset shows the compensated curve using a fitted scaling exponent to emphasize the power-law behavior.

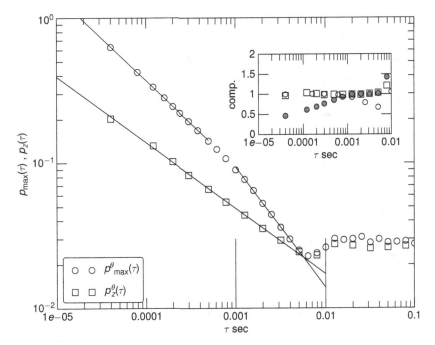

Figure 5.32 Measured maximum pdf $p_{max}(\tau)$ (○) and zero-cross $p_z(\tau)$ (□). A double power-law behavior is observed for $p_{max}(\tau)$ on the range $0.00004 < \tau < 0.0004$ sec, corresponding to a frequency range $2,500 < f < 25,000$ Hz, and $0.001 < \tau < 0.004$ sec, corresponding to $250 < f < 1,000$ Hz, respectively. The measured scaling exponents are 0.59 ± 0.01 and 0.83 ± 0.02. A single power-law behavior is observed for $p_z(\tau)$ on the range $0.00004 < \tau < 0.004$ sec, corresponding to a frequency range $250 < f < 25,000$ Hz, with a scaling exponent 0.45 ± 0.01. The inset shows the compensated curve to emphasize the observed scaling behavior.

$0.00004 < \tau < 0.0004$ seconds and $0.001 < \tau < 0.004$ seconds, respectively corresponding to the frequency range $2,500 < f < 25,000$ Hz and $250 < f < 1000$ Hz. The measured scaling exponent is 0.59 ± 0.01 and 0.83 ± 0.02. These scaling exponents are larger than the first-order nonintermittent KOC value $1/3$. The inset shows the compensated curve to emphasize the observed power-law behavior. The zero-crossing pdf $p_z^\theta(\tau)$ is also calculated, which is also shown in Figure 5.32 as □. Surprisingly, a nearly two decades power-law behavior is observed on the range $0.00004 < \tau < 0.004$ seconds, corresponding to a frequency range $250 < f < 25,000$ Hz. The corresponding scaling exponent is 0.45 ± 0.01. Note that for a fractional Brownian motion process with a scaling exponent H the corresponding zero-crossing scaling is to be H (see Equation [4.36]). Here, the observed zero-crossing scaling exponent deviates from the KOC scaling $1/3$.

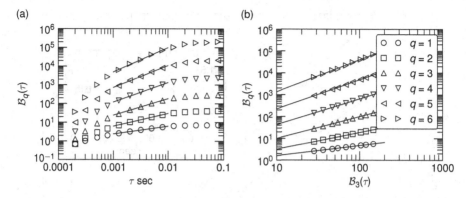

Figure 5.33 a) Measured maximum first-order detrended structure function $\mathcal{B}_q(\tau)$. Power-law behavior is observed on the range $0.001 < \tau < 0.01$ seconds, corresponding to a frequency range $100 < f < 1000$ Hz. b) The extended-self-similarity plot $\mathcal{B}_q(\tau)$ versus $\mathcal{B}_3(\tau)$ on the range $0.001 < \tau < 0.01$ seconds.

5.2.4 Detrended structure function

A first-order detrended structure function is performed by removing the first-order polynomial from a segment with a given separation scale. The detrended structure function $\mathcal{B}_q(\tau)$ is then calculated. Figure 5.33a shows the measured $\mathcal{B}_q(\tau)$. Power-law behavior is observed on the range $0.001 < \tau < 0.01$ seconds, corresponding to a frequency range $100 < f < 1000$ Hz. Note that this scaling range is the same as the one predicted by the Fourier power spectrum. This is because the detrending process constrains partially the influence of the ramp-cliff structures, which are now understood as energetic large-scale structures (Huang et al., 2010; Huang, 2014).

To retrieve the scaling exponent more accurately, the extended-self-similarity technique is also applied. The corresponding plot $\mathcal{B}_q(\tau)$ versus $\mathcal{B}_3(\tau)$ is shown in Figure 5.33b on the range $0.001 < \tau < 0.01$ seconds. The scaling exponent $\zeta_\theta(q)$ is then estimated on this scaling range. Figure 5.34a shows the measured scaling exponents. For comparison, the scaling exponent compiled by Schmitt (2005) is also shown as □. Note that the detrended scaling exponent is close to the one compiled by Schmitt (2005), while the relative one provided by the extended-self-similarity is slightly larger than the compiled one.

However, all these scaling exponents deviate from the lognormal formula with the intermittency parameter $\mu_\theta = 0.20$. To emphasize the multifractality, the singularity spectrum $f_\theta(\alpha_\theta)$ is then calculated. The experimental $f_\theta(\alpha_\theta)$ is shown in Figure 5.34b. The shape of the detrended-based singularity spectrum $f_\theta(\alpha_\theta)$ and the one provided by the extended-self-similarity are parallel when $\alpha_\theta \leq 0.4$. While the former one is almost the same as the compiled one, it is worth pointing out that the Reynolds number of the present data is $Re_\lambda = 250$. The results shown here

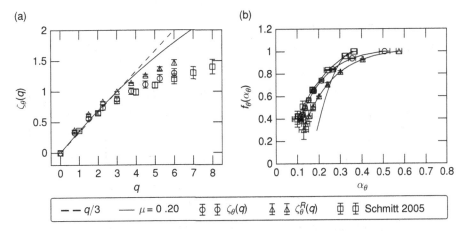

Figure 5.34 a) Measured detrended structure function scaling exponent $\zeta_\theta(q)$ (○) and the extended-self-similarity scaling exponent $\zeta_\theta^R(q)$ (△). For comparison, a scaling exponent $\zeta_\theta(q)$ compiled by Schmitt (2005) is also shown as □. A lognormal with an intermittent parameter $\mu_\theta = 0.20$ is illustrated as a solid line. b) The corresponding singularity spectrum $f_\theta(\alpha_\theta)$.

indicate that once the ramp-cliff structure is constrained, even partially, the retrieved scaling exponent and the singularity spectrum are close to the ones obtained from the large Reynolds number turbulent flows.

5.2.5 Wavelet leaders

Since the wavelet transform modulus maximum provides the same performance as the wavelet leaders (Huang et al., 2011a), we therefore consider here only the latter one. The temperature data is first divided into several segments, with 2^{14} data points each and 50% overlap. Figure 5.35a shows the measured $Z_q(\tau)$ on the range $0 \le q \le 6$. Power-law behavior is evident on the range $0.001 < \tau < 0.01$ sec, corresponding to a frequency range $100 < f < 1000$ Hz. The scaling exponent $\xi_\theta(q)$ is then estimated on this range. Figure 5.35b shows the measured singularity spectrum $f_\theta(\alpha_\theta)$ (○). For comparison, the compiled value is also shown as □. Visually, the wavelet leaders indicate more intermittency than the compiled one (Huang et al., 2011a). Note that this wavelet-based method is still influenced by ramp-cliff structures (Huang et al., 2011a).

5.2.6 Hilbert-Huang transform

Empirical mode decomposition

The temperature data is first divided into several segments with 2^{12} data points each and 50% overlap. The empirical mode decomposition is applied to each segment.

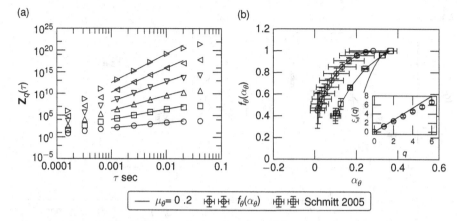

Figure 5.35 a) Experimental qth-order wavelet leaders-based moments $Z_q(\tau)$ on the range $0 \le q \le 6$. Power-law behavior is observed on the range $0.001 < \tau < 0.01$ sec. b) The corresponding singularity spectrum $f_\theta(\alpha_\theta)$. For comparison, the lognormal formula with an intermittency parameter $\mu_\theta = 0.20$ (solid line) and the compiled value ($\square$) are also shown. The inset shows the measured scaling exponent $\xi_\theta(q)$.

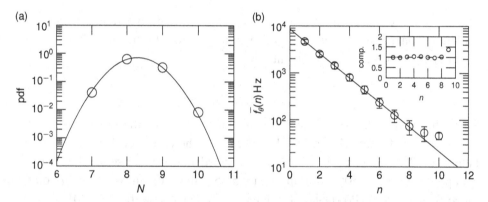

Figure 5.36 a) Experimental probability density function of the extracted number of intrinsic mode functions. For comparison, a Gaussian distribution is illustrated by a solid line. b) The corresponding mean frequency $\bar{f}_\theta(n)$ of each intrinsic mode function. An exponential law is observed with a scaling exponent 1.81 ± 0.03, indicating a dyadic-like filter bank property of the Empirical Mode Decomposition. The inset shows the compensated curve to emphasize the observed exponential law.

The number of the intrinsic mode function is in the range $N \in [7, 10]$. Figure 5.36a shows the measured pdf of the extracted N, in which a Gaussian distribution is illustrated as a solid line. It indicates that the experimental N satisfies very well the Gaussian distribution. The Fourier energy weighted mean frequency $\bar{f}(n)$ defined by Equation (5.6) is calculated and is shown in Figure 5.36b. An exponential law

5.2 Passive scalar

is observed on the range $n \leq 8$ with a scaling exponent 1.81 ± 0.03. This value is slightly smaller than the value 2 obtained for the turbulent velocity. It implies a quasi-dyadic-like filter bank. As mentioned previously, this value depends on several factors. One of them is the dynamics behind the analyzed data. The inset of Figure 5.36b shows the compensated curve to emphasize the observed exponential law.

Hilbert Spectral Analysis

We first check the second-order Hilbert marginal spectrum $\mathcal{L}_2^\theta(\omega)$, which is comparable with the Fourier power spectrum $E_\theta(f)$ (Huang et al., 2010, 2011a). Power-law behavior is visible on the range $100 < \omega < 1000\,\text{Hz}$, corresponding to a time scale range $0.001 < \tau < 0.01$ sec (Figure 5.37), and the corresponding scaling exponent $\xi_\theta(q) = 1.69 \pm 0.02$. Note that this value is very close to the nonintermittent KOC value 2/3.

We now turn to the qth-order moments $\mathcal{L}_q^\theta(\omega)$. Figure 5.38a shows the measured $\mathcal{L}_q^\theta(\omega)$ on the range $0 \leq q \leq 6$. Power-law behavior is observed on the range $100 < \omega < 1000\,\text{Hz}$. For display clarity, these curves have been vertical shifted. The scaling exponent $\xi_\theta(q)$ is then estimated on this scaling range. The zeroth-order scaling exponent is found to be $\xi_\theta(0) = 0.90 \pm 0.01$. To emphasize the

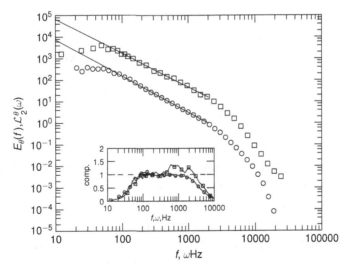

Figure 5.37 Comparison of the second-order Hilbert marginal spectrum $\mathcal{L}_2^\theta(\omega)$ (○) and the Fourier power spectrum $E_\theta(f)$ (□). Power-law behavior is observed on the range $100 < f < 1000\,\text{Hz}$ with scaling exponent $\xi_\theta(2) = 1.69 \pm 0.02$ and 1.56 ± 0.03, respectively for the Hilbert and Fourier approaches. The inset shows the compensated curve using the nonintermittent KOC value 2/3.

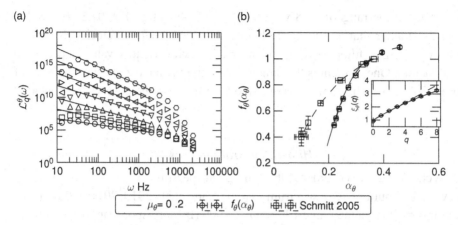

Figure 5.38 a) Measured qth-order Hilbert marginal spectrum $\mathcal{L}_q^\theta(\omega)$ on the range $0 \le q \le 6$. Power-law behavior is observed on the range $100 < \omega < 1000\,\text{Hz}$. b) Experimental singularity spectrum $f_\theta(\alpha_\theta)$. For comparison, the lognormal formula with an intermittency parameter $\mu_\theta = 0.20$ (solid line) and the compiled value ($\square$) are also shown. The inset shows the measured scaling exponent $\xi_\theta(q)$.

multifractality, the singularity spectrum is defined via the Legendre transform as,

$$\alpha_\theta = \frac{d\xi_\theta(q)}{dq}, f_\theta(\alpha_\theta) = \min_q \{\alpha_\theta q - \xi_\theta + 2\} \qquad (5.10)$$

Figure 5.38b shows the measured singularity spectrum $f_\theta(\alpha_\theta)$ ($\bigcirc$). For comparison, the compiled one is also shown as $\square$. Note that once the ramp-cliff structures are constrained, the measured singularity spectrum agrees very well with the lognormal formula with $\mu_\theta = 0.20$. Thus the result shown here indicates that the passive scalar field might be less intermittent than what was previously believed (Warhaft, 2000).

5.2.7 General remarks on passive scalar turbulence

As mentioned above, the passive scalar field was occasionally treated as a simple complementary of the velocity field since it is mainly advected by the velocity field. Later, experiments and numerical simulations confirmed that due to the presence of ramp-cliff structures, it has its own statistics, showing a more intermittent property than the embedded velocity field (Warhaft, 2000). The ramp-cliff structures are now understood as energetic large-scales, which possesses gradually decreasing "ramps" and rapidly increasing "cliffs." Traditional methods are strongly influenced by such structures. The terminology of "traditional methods" includes the classical structure function analysis, detrended fluctuation analysis, wavelet transform modulus maximum, wavelet leaders, etc. This energetic large-scale effect is regarded as infrared

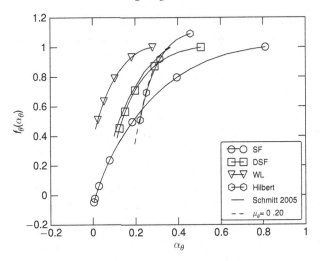

Figure 5.39 The singularity spectra $f_\theta(\alpha_\theta)$ provided by different approaches.

effect (Huang et al., 2013). Traditional approaches either mix the scale information or require high-order harmonics to represent the nonlinear and nonstationary events, for example, ramp-cliff structures, or vortex trapping events. The corresponding statistics are then biased (Huang et al., 2011a). Once this infrared effect is constrained/removed, the retrieved scaling exponent and singularity spectrum then approaches to the one of turbulent velocity. We finally reproduce in Figure 5.39 the singularity spectra $f_\theta(\alpha_\theta)$ provided by different approaches.

Note that different approaches might imply different intensity of intermittency. Therefore, a suitable method should be carefully chosen. Moreover, it is worth pointing out that the statistics of passive scalar might also depend on the Schmidt number, which is defined as,

$$Sc = \frac{\nu}{D} = \frac{\text{viscous diffusion rate}}{\text{molecular (mass) diffusion rate}} \quad (5.11)$$

in which ν is the kinematic viscosity, and D is the mass diffusivity. Therefore, the statement "the passive scalar turbulence might be less intermittent than what is previously believed" should be checked systematically for different Sc since different values of Sc in the passive scalar field might be different.

5.3 The Lagrangian Turbulence

The Lagrangian view of turbulence, in which the fluid parcel is tracked experimentally or numerically (Falkovich et al., 2001; Yeung, 2002; Toschi and Bodenschatz, 2009), provides an adequate description of turbulent phenomena, such as relative dispersion, air pollution, etc. With respect to the Kolmogorov 1941 theory, the Lagrangian version predicts a scaling behavior in the temporal domain,

$$S_q^L(\tau) = \langle \Delta_\tau V(t)^q \rangle \sim \epsilon^{q/2} \tau^{q/2} \tag{5.12}$$

in which ϵ is the mean energy dissipation rate, and τ is the separation time scale, which is lying in the so-called inertial range. This was first proposed in 1944 by Landau, in the Russian version of his book (see Monin and Yaglom [1971]). With the increasing ability of experiment techniques and numerical computation capabilities, several attempts have been made to verify the above so-called Kolmogorov-Landau theory (Chevillard et al., 2003; Biferale et al., 2004; Xu et al., 2006a,b; Arnéodo et al., 2008). However, the above Lagrangian velocity structure function is strongly influenced by the presence of vortex trapping events (see example shown in Figure 5.47). It is thus difficult to validate Equation (5.12) even for the case $q = 2$ (Sawford and Yeung, 2011; Falkovich et al., 2012). Note that the vortex trapping possesses a typical time scale around $3 \sim 5\tau_\eta$ and corresponds to a strong dissipation event. Therefore, a careful treatment of the Lagrangian velocity is required to handle this special event, known as ultraviolet effect (Huang et al., 2013).

We consider below a dataset composed by Lagrangian velocity trajectories in a three-dimensional homogeneous and isotropic turbulent flow obtained from a 2048^3 direct numerical simulation with a Reynolds number $Re_\lambda = 400$. We recall briefly some key parameters of this database. There are $\sim 2 \cdot 10^5$ fluid tracer trajectories, each composed of $N = 4720$ time sampling saved every $0.1\tau_\eta$ time units, in which τ_η is the Kolmogorov time scale. Hence, we can access time scales in the range $0.1 < \tau/\tau_\eta < 236$. An inertial range $0.01 < \omega\tau_\eta < 0.1$ (resp. $10 < \tau/\tau_\eta < 100$) has been reported for this database by using a Hilbert-based methodology (Huang et al., 2013). We therefore focus on this inertial range in the following analysis. The details of this database can been found in Benzi et al. (2009).

5.3.1 Fourier power spectrum and second-order structure functions

Figure 5.40a shows the measured Lagrangian Fourier power spectrum $E_L(f)$, in which the inset shows a compensated curve using the nonintermittent Kolmogorov-Landau value 2. Power-law behavior is observed on the range $0.01 < f\tau_\eta < 0.1$, corresponding to a time scale range $10 < \tau/\tau_\eta < 100$. The fitted scaling exponent is found to be 1.88 ± 0.02, which is slightly smaller than the nonintermittent Kolmogorov-Landau value 2. Note that the Fourier power spectrum is expected to scale as $E_L(f) \sim \epsilon f^{-2}$. Therefore, if one takes into account the refined similarity hypothesis, there is no intermittent correction involved. The violation of the "-2"

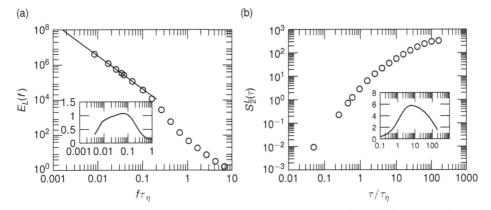

Figure 5.40 a) Experimental Fourier power spectrum $E_L(f)$ of the Lagrangian $V_x(t)$. Power-law behavior is observed on the range $0.01 < f\tau_\eta < 0.1$ with a scaling exponent 1.88 ± 0.02. The inset shows the compensated curve using the nonintermittent Kolmogorov-Landau value 2. b) Measured second-order Lagrangian velocity structure function $S_2^L(\tau)$. The inset shows the compensated curve by using the nonintermittent Kolmogorov-Landau value 1.

scaling is now understood as partially an effect of finite Reynolds number, and partially the nonlinear contamination of vortex trapping events (Huang et al., 2013).

Figure 5.40b shows the measured second-order Lagrangian structure function $S_2^L(\tau)$, in which the inset shows the compensated curve using the Kolmogorov-Landau theory value "1". The shape of $S_2^L(\tau)$ and the compensated curve $S_2^L(\tau)\tau^{-1}$ are similar with the ones reported in other references (Sawford and Yeung, 2011; Falkovich et al., 2012). There is no clear inertial range observed. This has been previously interpreted as an effect of the finite Reynolds number of turbulent flows. However, with the increase of the Reynolds number both numerically and experimentally, there is no tendency to retrieve the Kolmogorov-Landau scaling (Sawford and Yeung, 2011; Falkovich et al., 2012). This thus cast doubts on the correctness and accuracy of the present phenomenological models. Note that several theoretical works have attempted to have a better understanding of the Lagrangian statistics in the sense of the anomalous scaling (Schmitt, 2006; Beck, 2007; He, 2011; Zybin et al., 2008).

Now, the difficulty to identify the Kolmogorov-Landau scaling is recognized as coming from two effects. The first one is the mixture of scales due to the global property of statistical methodologies. It is usually manifested as the contamination of large-scale structures (infrared effect). The second one is the influence of vortex-trapping events within the dissipation range (ultraviolet effect). Note that vortex trapping events are visible at the pdf tail of the Lagrangian velocity increment. Therefore, the high-order statistics might be dominated by it. For example, Xu et al.

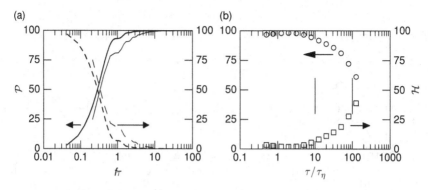

Figure 5.41 a) Measured cumulative function $\mathcal{P}(\tau,f)$ (solid line) and the corresponding cocumulative function $\mathcal{H}(\tau,f)$ (dashed line) at two time scales $\tau/\tau_\eta = 10$ (thick solid line) and $\tau/\tau_\eta = 50$ (thin solid line), corresponding $f\tau_\eta = 0.1$ and 0.02 in the inertial range. b) The corresponding $\mathcal{P}_1(\tau)$ (○) and $\mathcal{H}_1(\tau)$ (□). The inertial range $10 < \tau/\tau_\eta < 100$ predicted by the Fourier power spectrum is illustrated by a solid line.

(2006a) performed an experimental study of the high-order Lagrangian velocity structure-function and estimated the extended-self-similarity scaling exponent on the range $3 \sim 6\tau_\eta$. They found a saturation phenomenon for moments $q \geq 4$. The inertial range $3 \sim 6\tau_\eta$ they chose is still in the dissipation range. Therefore, it is likely that the saturation reveals the vortex trapping dynamics. We will turn back again to this topic in Section 5.3.4 and provide general comments in Section 5.6.

To have a better understanding of the scale contribution to the second-order Lagrangian velocity structure, we calculate the cumulative function $\mathcal{P}(\tau,f)$ by using the experimental Fourier spectrum $E_L(f)$. Figure 5.41a shows the measured $\mathcal{P}(\tau,f)$ for two separation scales $\tau/\tau_\eta = 10$ (thin solid line) and $\tau/\tau_\eta = 50$ (thick solid line), corresponding to $f/\tau_\eta = 0.01$ and 0.02 in the inertial range. The cocumulative function $\mathcal{H}(\tau,f)$ is also shown as a dashed line. Visually, the large-scale part $f\tau \leq 1$ has more than 75% contribution. This is because the measured $\mathcal{P}(\tau,f)$ is increasing with β, the slope of the Fourier spectrum, $E(f) \sim f^{-\beta}$, see also Figure 4.3 and discussion in section 4.2.4. Or in other words, the steeper the Fourier spectrum, the larger the large-scale contamination.

Note that large-scale structures are usually anisotropic, which might be associated with flow geometry, forcing, etc. Such anisotropy violates the theory requirement, such as local homogeneity and isotropy. Figure 5.41b shows the corresponding $\mathcal{P}_1(\tau)$ (○) and $\mathcal{H}_1(f)$ (□), in which the inertial range is indicated by a solid line. It confirms the observation in Figure 5.41a that in the inertial range the large-scale part has more than 50% contribution.

5.3.2 High-order structure functions and extended-self-similarity

Figure 5.42a shows the measured high-order Lagrangian velocity structure functions $S_q^L(\tau)$, in which the inertial range $10 < \tau/\tau_\eta < 100$ is demonstrated by a solid line. As mentioned above due to the presence of vortex trapping events and the effect of large-scale contamination, there is no clear power-law behavior. With the help of extended-self-similarity, the power-law behavior and the associated scaling exponent could be identified. Figure 5.42b shows the extended-self-similarity plot $S_q^L(\tau)$ versus $S_2^L(\tau)$ on the range $3 \le \tau/\tau_\eta \le 100$. Visually, a single power-law behavior is observed for low-order moments, while for high-order ones, a dual power-law behavior is visible respectively on the range $3 \le \tau/\tau_\eta \le 30$ (open symbols) and $30 \le \tau/\tau_\eta \le 100$ (closed symbols).

The relative scaling exponent $\zeta_L(q)$ is then estimated on these two scaling ranges. Figure 5.43a shows the measured scaling exponent $\zeta_L(q)$ obtained from the above-mentioned scaling ranges, denoted as $\zeta_L^1(q)$ for the former scaling range and $\zeta_L^2(q)$ for the later one. For comparison, the value provided by Xu et al. (2006a) ($\triangle$) and the multifractal model of Biferale et al. (2004) (thick solid line) are also shown. The measured scaling exponent $\zeta_L^1(q)$ agrees very well with the one provided by Xu et al. (2006a), since the first scaling range is on the range $3 \sim 30\, \tau/\tau_\eta$, in which the vortex trapping event is involved $3 \sim 5\, \tau/\tau_\eta$. It thus implies a more intermittent inertial range dynamics if one takes this range as the inertial range (Chevillard et al., 2003; Beck, 2007; Zybin et al., 2008). The measured $\zeta_L^2(q)$ agrees very well with the multifractal model of Biferale et al. (2004) since the second scaling range is

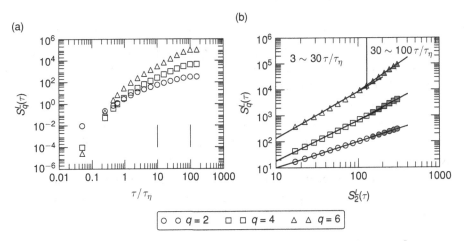

Figure 5.42 a) Experimental high-order Lagrangian structure functions $S_q^L(\tau)$, in which the inertial range predicted by the Fourier power spectrum is illustrated by a solid line. b) Extended-self-similarity plot $S_q^L(\tau)$ versus $S_2^L(\tau)$ on the range $3 \le \tau/\tau_\eta \le 100$. The vertical solid line indicates a separation scale $\tau/\tau_\eta = 30$.

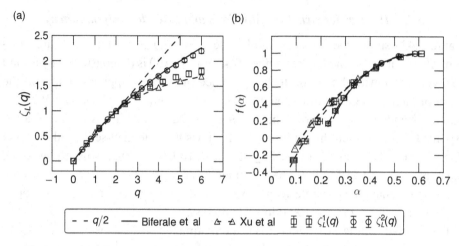

Figure 5.43 a) Measured extended-self-similarity scaling exponent $\zeta_L(q)$. For comparison, the value predicted by the multifractal model of Biferale et al. (2004) (solid line) and the value from the experiment (denoted as Xu et al., $\triangle$) are also shown. b) The corresponding singularity spectrum $f(\alpha)$.

lying in the inertial range $10 < \tau/\tau_\eta < 100$. To emphasize this observation, we calculate the singularity spectrum $f(\alpha)$. Figure 5.43b displays the measured $f(\alpha)$. It confirms the above statements.

5.3.3 Hilbert-Huang transform

Empirical mode decomposition

The empirical mode decomposition is applied to the Lagrangian velocity of each tracer particle. The intrinsic mode function is then extracted with a number of modes $N \in [4, 9]$. Figure 5.44a shows the measured probability density function $p(N)$, in which the solid line indicates a Gaussian distribution with the same mean value and variance. The measured $p(N)$ agrees well with the Gaussian distribution. The mean frequency $\bar{f}(n)$ of each mode is then estimated. Figure 5.44b shows the measured $\bar{f}(n)$ for the first six modes. An exponential-law is observed with a measured scaling exponent $\gamma = 1.75 \pm 0.05$, indicating a dyadic-like filter bank. Note that the value γ not only depends on the sifting number as reported by Wu and Huang (2010), but also depends on the process analyzed. For example, this value is found to be $\gamma = 1.81$ for the passive scalar, and $\gamma = 1.7$ for the wind power (Calif et al., 2013).

High-order Hilbert moments

After retrieving modes from each particle, the Hilbert spectral analysis is then performed. The qth-order Hilbert moments $\mathcal{L}_q(\omega)$ are then calculated. Figure 5.45

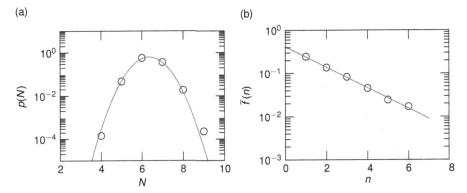

Figure 5.44 a) Experimental probability density function of the number of extracted intrinsic mode functions, in which the Gaussian distribution is illustrated by a solid line. b) Measured mean frequency $\bar{f}(n)$ for the first six modes. An exponential-law, e.g., $\bar{f}(n) \sim \gamma^{-n}$ is observed with a scaling exponent $\gamma = 1.75 \pm 0.05$, indicating a dyadic-like filter bank of the empirical mode decomposition.

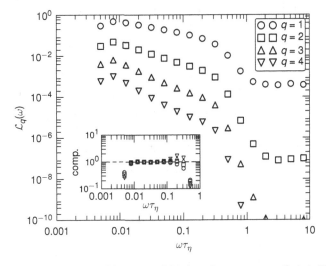

Figure 5.45 Experimental Hilbert-based high-order moments $\mathcal{L}_q(\omega)$. Power-law behavior is observed on the range $0.01 < \omega\tau_\eta < 0.1$ for all q considered here. The inset shows the compensated curve using the fitted slope to emphasize the observed scaling behavior.

shows the measured $\mathcal{L}_q(\omega)$. Power-law behavior is observed on the range $0.01 < \omega\tau_\eta < 0.1$, corresponding to a time scale range $10 < \tau/\tau_\eta < 100$ (Huang et al., 2013). We then take this range as the inertial range and fit the scaling exponent using a least-square fitting algorithm. Note that the vortex trapping event with the typical time scale $3 \sim 5\,\tau_\eta$ is excluded in this inertial range. The corresponding second-order scaling exponent is $\zeta_L(2) \simeq 1.03$, which agrees well with the

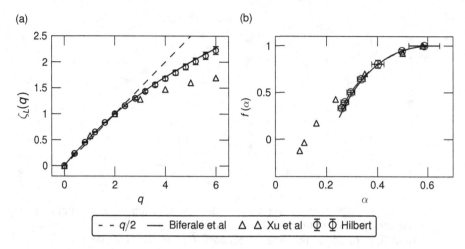

Figure 5.46 a) Experimental Hilbert-based scaling exponent $\zeta_L(q)$ ($\bigcirc$), in which the errorbar indicates 95% fitting confidence interval on the range $0.01 < \omega\tau_\eta < 0.1$. b) The corresponding singularity spectrum $f(\alpha)$ versus α. For comparison, the experiment value provided by Xu et al. (2006a) is also shown as $\triangle$.

Kolmogorov-Landau prediction. To emphasize the observed power-law behavior, the compensated curve using a fitted scaling exponent is then shown as the inset in Figure 5.45. A clear plateau is evident, confirming the existence of the Lagrangian inertial range on the range $0.01 < \omega\tau_\eta < 0.1$ (Huang et al., 2013). Note that the scaling behavior is here identified without the help of the extended-self-similarity, which is widely used in the Lagrangian scaling analysis (Xu et al., 2006a; Arnéodo et al., 2008; Berg et al., 2009). The scaling exponent $\zeta_L(q)$ is then estimated on this inertial range.

Figure 5.46 a shows the retrieved Hilbert based scaling exponent $\zeta_L(q)$ ($\bigcirc$), in which the multifractal formula is shown as a solid line. Note that the errorbar indicates a 95% fitting confidence interval. For comparison, the experimental value provided by Xu et al. (2006a) is also shown as $\triangle$. Visually, the measured scaling exponent agrees perfectly with the multifractal model of Biferale et al. (2004), and disagrees with the experimental data by Xu et al. (2006a) when $q > 3$, which indicates a more intermittent Lagrangian velocity (Chevillard et al., 2003). Figure 5.46 b shows the corresponding singularity spectrum $f(\alpha)$ versus α. The experiment curve provided by Xu et al. (2006a) is much wider than the Hilbert-based one.

5.3.4 General remarks on the Lagrangian view of turbulence

We now turn to the vortex trapping events, which cause the difficulty in the analysis of the Lagrangian velocity. Figure 5.47a shows a typical vortex trapping event

5.3 The Lagrangian Turbulence

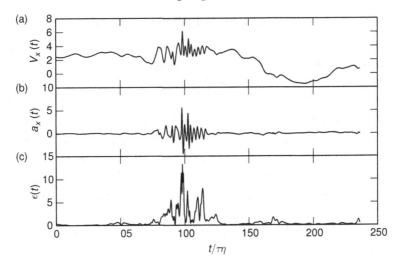

Figure 5.47 A vortex trapping event along a Lagrangian trajectory: a) Lagrangian velocity $V_x(t)$, b) the corresponding acceleration $a_x(t)$, and c) the energy dissipation rate $\epsilon(t)$ along this trajectory. The vortex trapping event is visible on the time span $80 < t/\tau_\eta < 120$ with a typical time scale $3 \sim 5\,\tau_\eta$.

(see also Figure 1.5). The observed vortex trapping event covers a time span $80 < \tau/\tau_\eta < 120$ with 11 oscillations. It corresponds to a typical time scale of around $3 \sim 5\,\tau_\eta$ (Toschi et al., 2005; Bec et al., 2006; Huang et al., 2013). During the vortex trapping event, the fluid parcel is strongly accelerated by the vortex filament and thus suffers a high intensity dissipation (see Figures 5.47b and c). It implies that this special event belongs to the dissipation range (Toschi et al., 2005; Bec et al., 2006; Huang et al., 2013). Therefore, the right choice of the inertial range should exclude vortex trapping events. More precisely, the correct inertial range should at least be above $5\,\tau_\eta$.

We note that previously at least two ranges have been chosen as the Lagrangian inertial range. They are $3 < \tau/\tau_\eta < 6$ used by Xu et al. (2006a) in high intensity turbulent experiments, and $10 < \tau/\tau_\eta < 100$ used by Biferale et al. (2004) and Huang et al. (2013) in a high resolution direct numerical simulation database. The vortex trapping is included in the former one. This is thus a possible explanation of the saturation phenomenon when $q > 4$. As shown above, once the vortex trapping event is excluded, the second-order statistics show a clear Kolmogorov-Landau scaling behavior with a slope -1, and the retrieved scaling exponent $\zeta_L(q)$ agrees very well with the multifractal formula. All these results are compatible with the classification of the vortex trapping effect as an ultraviolet effect in the dissipation range (Huang et al., 2013).

5.4 Rayleigh-Bénard turbulent convection

In Rayleigh-Bénard (RB) convection cells, the flow is dominated by buoyant forces mainly via thermal plumes. According to Bolgiano and Obukhov arguments, above the Bolgiano length scale ℓ_B (see definition below), instead of the Kolmogorov-Obukhov-Corrsin (KOC) scaling $S_q^\theta(r) \sim r^{q/3}$ for the passive scalar, the Bolgiano-Obukhov (BO59) scaling $S_q^\theta(r) \sim r^{q/5}$ for the active scalar is expected for the temperature in which r is the separation length scale and lies in the inertial range (Bolgiano, 1959; Obukhov, 1959; Lohse and Xia, 2010). However, there exist several difficulties in identifying the BO59 scaling for the qth-order structure functions in both temporal and spatial domains (Lohse and Xia, 2010). These difficulties are recognized as the anisotropy and the inhomogeneity of the flow (Kunnen et al., 2008; Zhou and Xia, 2011), and as a consequence the BO59 argument may fail. Other reasons for such failure include too low Rayleigh numbers, intermittent corrections, the presence of a shear, and the lack of clear separation between Bolgiano length scale ℓ_B and the height of the cell L, etc. (Lohse and Xia, 2010). Thus, whether or not the BO59 scaling exists in a turbulent RB system, it is still a major challenge. For the small-scale statistics in a RB convection cell, we refer the interested reader to the review paper by Lohse and Xia (2010).

The classical structure functions analysis is the most often used method to extract the small-scale scaling (Kunnen et al., 2008; Lohse and Xia, 2010; Zhou and Xia, 2011), which is written as (we work in temporal domain and focus on the temperature θ)

$$S_q(\tau) \equiv \langle |\Delta\theta_\tau(t)|^q \rangle \sim \tau^{\zeta_\theta(q)} \tag{5.13}$$

in which $\Delta\theta_\tau(t) = \theta(t+\tau) - \theta(t)$ is the temperature increment, and τ the separation scale (Frisch, 1995; Warhaft, 2000; Lohse and Xia, 2010). Above the Bolgiano time scale t_B (see definition below) one expects the BO59 scaling for the active scalar

$$\zeta_\theta(q) = q/5 \tag{5.14}$$

in which the intermittent correction is ignored. If the temperature fluctuation acts as a passive scalar, one has the KOC scaling

$$\zeta_\theta(q) = q/3 \tag{5.15}$$

where the intermittent correction is ignored as well (Warhaft, 2000; Lohse and Xia, 2010). We note that several researchers have reported BO59-like structure-functions for the temperature using temporal statistics or directly spatial statistics (Cioni et al., 1995; Zhou and Xia, 2001; Kunnen et al., 2008). However, even in Fourier space the scaling range is quite short, usually less than one decade

5.4 Rayleigh-Bénard turbulent convection 145

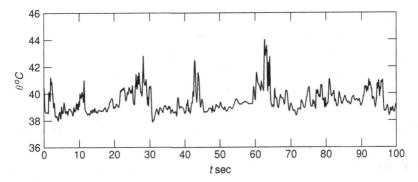

Figure 5.48 A 100 sec of temperature measured in the near side wall of the Rayleigh-Bénard convection cell with a Rayleigh number $Ra = 1.31 \times 10^{10}$. A thermal plume, energetic structure is visible.

(Ashkenazi and Steinberg, 1999; Niemela et al., 2000; Zhou and Xia, 2001; He and Tong, 2011).

Similarly with the case of the passive scalar turbulence (see Section 5.2), a typical energetic structure, namely thermal plume, is observed in the temperature field in the near side wall region. Figure 5.48 shows a 100 sec portion of the experiment data collected by Professor Xia at the Chinese University of Hong Kong. High intensity events are observed when the thermal plume passes by. They correspond to the unique thermal structure in the turbulent RB convection. The classical structure function might be influenced by this structure.

Another aspect of the RB system is the presence of the so-called large-scale-circulation (LSC), which is observed in numerous studies, including experiments, numerical simulations, and theoretical analysis, for example, Hartlep et al. (2003); Xi et al. (2004); Brown et al. (2005); Brown and Ahlers (2007); Xi and Xia (2007); Zhou et al. (2009), to quote a few. The LSC is a self-organized large-scale structure and driven by buoyant forces, mainly via thermal plumes initialed from the thermal boundary layer. It certainly belongs to the energetic large-scale structures, which obviously is nonlinear and nonstationary in both temporal and spatial domains (Brown and Ahlers, 2007). Thus it might suffer from the same problem as the passive scalar. Indeed, the plume structure has been recognized as a cliff-ramp-like structure, in which a sharp cup is followed by a long tail (Grossmann and Lohse, 2004; Zhou et al., 2007). For interested readers, we refer to the recent review papers by Ahlers et al. (2009) and by Lohse and Xia (2010).

We now present the data used in this section. The experiments were performed by Prof. Xia's group at the Chinese University of Hong Kong. The details of the experiments have been described elsewhere (Shang et al., 2003, 2004, 2008). We describe briefly the experiment below. The temperature measurements were carried out in a cylindrical cell with upper and lower copper plates and a Plexiglas sidewall. The inner diameter of the cell is $D = 19.0$ cm and the height is $L = 19.6$ cm, thus its aspect ratio is $\Gamma \equiv D/L \simeq 1$. Water was used as the working fluid and measurements were made at Rayleigh number $Ra \equiv \beta g \Delta L^3/\nu\kappa = 9.5 \times 10^9$ and 1.31×10^{10} with g being the gravitational acceleration, Δ the temperature difference across the fluid layer, and β, ν, and κ being, respectively, the thermal expansion coefficient, the kinematic viscosity, and the thermal diffusivity of the working fluid (water). During the experiments, the entire cell was placed inside a thermostat box whose temperature matched the mean temperature of the bulk fluid, which was kept at $\sim 40\,°C$, corresponding to a Prandtl number $Pr = 4.4$. The local temperature was measured at 8 mm from the sidewall at mid-height using a small thermistor of 0.2 mm diameter and 15 ms time constant. Typically, each measurement of temperature lasted 20 hours or longer with a sampling frequency of 64 Hz, ensuring that the statistical averages converged.

We note that the Bolgiano length scale can be estimated as

$$\ell_B \equiv \frac{\epsilon_\nu^{5/4}}{\epsilon_T^{3/4}(g\alpha)^{3/2}} = \frac{Nu^{1/2}L}{(RaPr)^{1/4}} \qquad (5.16)$$

where ϵ_ν and ϵ_T are the kinematic and thermal dissipation rates, and Nu is the key response of the RB system, and the Nusselt number, and the dimensionless heat flux (Siggia, 1994; Ahlers et al., 2009; Lohse and Xia, 2010). The counterpart of the Bolgiano length scale ℓ_B, the Bolgiano time scale in time domain thus can be estimated as

$$t_B \equiv \frac{\ell_B}{U} = \frac{Nu^{1/2}t_0}{(RaPr)^{1/4}} \qquad (5.17)$$

in which t_0 is the LSC turnover time (Siggia, 1994; Zhou and Xia, 2008). For the present experiments, t_B is found to be ~ 1 sec (Zhou and Xia, 2001, 2008). We thus expect the BO59 scaling in the near sidewall region, in which the flow is driven by buoyancy in the vertical direction via thermal plumes and above a certain scale, for example, t_b, buoyancy effects indeed become predominant, at least in the time domain (Ching et al., 2004; Zhou and Xia, 2008).

To relate our analysis results obtained in the temporal domain to theoretical predictions in the spatial domain, we invoke the elliptic model (He and Zhang, 2006; Zhao and He, 2009), which was advanced based on a systematic second-order Taylor-series expansion of the space-time velocity correlation functions $C(r,\tau)$. Indeed, several experimental studies in turbulent RB convection have shown that the scaling exponents measured in the temporal domain are the same as those obtained in the spatial domain if the elliptic model is invoked (He et al., 2010; He and Tong,

2011; Zhou et al., 2011) and some recent *Re*-measurements were also based on the elliptic model (He et al., 2012).

5.4.1 Fourier power spectrum and second-order structure function

Figure 5.48 shows a 100 sec portion of the collected temperature data, showing the typical thermal plumes. Figure 5.49 a shows the Fourier power spectral density function $E_\theta(f)$ in log-log plot. Power-law behavior is observed on the frequency range $0.1 < f < 1$ Hz, corresponding to the time scale $1 < \tau < 10$ sec (Huang et al., 2011b). The corresponding scaling exponent is found to be around $\simeq 1.20$. The LSC is found around $f_L \simeq 0.03$ Hz, corresponding to a time scale $\simeq 33$ sec. Figure 5.49 b shows the spectrum in a semi-log plot, in which the inset shows the autocorrelation function $\rho_\theta(\tau)$ with $0 < \tau < 50$ sec. As mentioned in the previous section the LSC oscillation is nonlinear and nonstationary in both spatial and temporal domains. Thus the f_L is a mean frequency of LSCs. We may consider the frequency range between two peaks $0.03 < f < 0.06$ Hz, corresponding to $15 < \tau < 33$ sec, as the range of the large-scale structures.

Figure 5.50 a shows the measured second-order structure function $S_2^\theta(\tau)$ for the case $Ra = 1.31 \times 10^{10}$. Power-law behavior is observed on the range $1 < \tau < 10$ sec with a scaling exponent $\zeta_\theta(2) = 0.27 \pm 0.03$. It is significantly smaller than the BO59 value 2/5. As discussed above, this experiment scaling exponent is biased

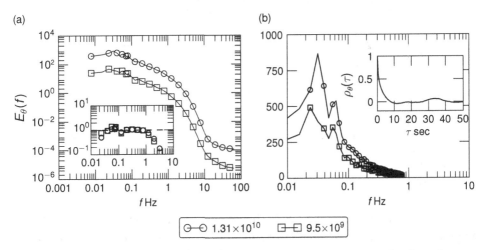

Figure 5.49 a) Experimental Fourier power spectrum $E_\theta(f)$ in a log-log plot. Power-law behavior is observed on the range $0.1 < f < 1$ Hz with a fitted scaling exponent $\simeq 1.20$. The inset shows a compensated curve using the fitted scaling exponent. b) Semi-log plot of $E_\theta(f)$ to emphasize the large-scale circulation with a typical frequency $f_L \simeq 0.03$ Hz, corresponding to a time scale $T_L = 33$ sec. The inset shows the measured autocorrelation function $\rho_\theta(\tau)$.

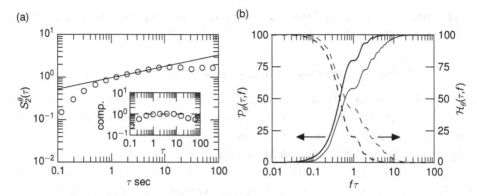

Figure 5.50 a) Experimental second-order structure function $S_2^\theta(\tau)$ for the case $Ra = 1.31 \times 10^{10}$. Power-law behavior is observed on the range $1 < \tau < 10$ sec with a fitted scaling exponent $\zeta_\theta(2) = 0.27 \pm 0.03$. The inset shows a compensated curve using the fitted scaling exponent. b) Measured cumulative function $\mathcal{P}_\theta(\tau,f)$ (solid line) and cocumulative function $\mathcal{H}_\theta(\tau,f)$ (dashed line) for separation time scale $\tau = 1$ sec (thick solid line) and $\tau = 5$ sec (thin solid line).

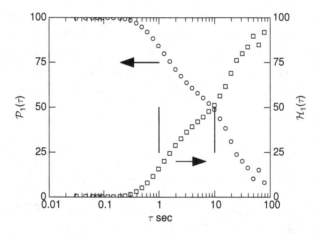

Figure 5.51 Experimental cumulative function $\mathcal{P}_1(\tau)$ ($\bigcirc$) and co-cumulative function $\mathcal{H}_1(\tau)$ ($\square$). The expected power-law range $1 < \tau < 10$ sec is illustrated by vertical lines.

by the thermal plumes and the LSC. The cumulative function $\mathcal{P}_\theta(\tau,f)$ (solid line) and the corresponding co-cumulative function $\mathcal{H}_\theta(\tau,f)$ are then estimated for the separation time scales $\tau = 1$ sec and τ sec lying in the power-law range. They are shown in Figure 5.50b. Visually, it implies that the measured structure function is strongly biased by the large-scale part, for example, $f\tau < 1$.

Figure 5.51 shows the measured $\mathcal{P}_1(\tau)$ ($\bigcirc$) and $\mathcal{H}_1(\tau)$ ($\square$), in which the power-law range is indicated by vertical lines. It suggests that more than 50% of the

5.4.2 High-order structure functions

High-order structure functions $S_q^\theta(\tau)$ are then estimated for $0 < q < 6$. Figure 5.52 shows the measured $S_q^\theta(q)$ for the case a) $Ra = 9.5 \times 10^9$, and b) $Ra = 1.31 \times 10^{10}$. Power-law behavior is observed on the range $1 < \tau < 10$ sec for all q considered here. The scaling exponent $\zeta_\theta(q)$ is then estimated on this range using a least square fitting algorithm. Figure 5.53a shows the measured $\zeta_\theta(q)$, in which the

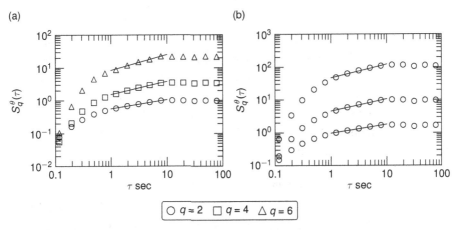

Figure 5.52 Experimental high-order structure functions $S_q^\theta(\tau)$: a) $Ra = 1.31 \times 10^{10}$, and b) $Ra = 9.5 \times 10^9$. Power-law behavior is observed on the range $1 < \tau < 10$ sec.

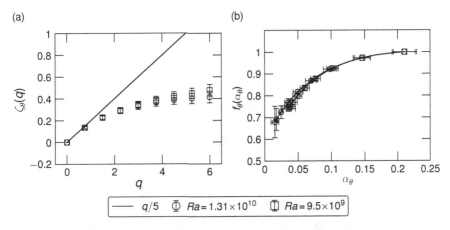

Figure 5.53 a) Experimental scaling exponent $\zeta_\theta(q)$. b) The corresponding singularity spectrum $f_\theta(\alpha_\theta)$ versus α_θ.

BO59 scaling is shown as a solid line. The measured $\zeta_\theta(q)$ deviates significantly from the BO59 value, indicating a strongly intermittent active scalar field. Note that the measured scaling exponent $\zeta_\theta(q)$ tends to be saturated for a very large q. Zhou and Xia (2002) reported a saturation value $\zeta_\theta(\infty) = 0.8$. To emphasize the intermittency, the singularity spectrum $f_\theta(\alpha_\theta)$ is estimated. Figure 5.53b shows the measured $f_\theta(\alpha_\theta)$. A wide range of α_θ and $f_\theta(\alpha_\theta)$ confirms the strong intermittency of this temperature time series.

5.4.3 Increments-based approaches

Autocorrelation function of increments

Figure 5.54 shows the measured Γ_θ for the case $Ra = 9.5 \times 10^9$ (○) and 1.31×10^{10} (□). Power-law behavior is observed on the range $1 < \tau < 10$ sec. The measured scaling exponent is 0.42 ± 0.02 for both cases. To emphasize this power-law behavior, the inset shows the compensated curve using the fitted scaling exponent. A clear plateau is observed, confirming the existence of the power-law behavior. Note that this measured exponent 0.42 is close to the second-order BO59 value 2/5. This is because the method of the autocorrelation function of increments can partially reduce the influence of the large-scale circulation.

Maximum pdf scaling of increments

Figure 5.55 shows the measured $p_{\max}^\theta(\tau)$. Power-law behavior is observed on the range $1 < \tau < 10$ sec with a scaling exponent 0.33 ± 0.02 (Huang et al., 2011b).

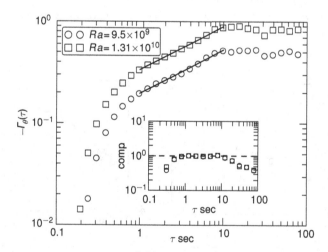

Figure 5.54 Experimental $\Gamma_\theta(\tau)$. Power-law behavior is observed on the range $1 < \tau < 10$ sec with a scaling exponent 0.42 ± 0.02. The inset shows the compensated curve using the fitted scaling exponent.

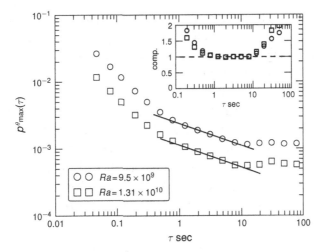

Figure 5.55 Experimental $p^\theta_{\max}(\tau)$. Power-law behavior is observed on the range $1 < \tau < 10$ sec with a scaling exponent 0.33 ± 0.02. The inset shows the compensated curve using the fitted scaling exponent.

This measured scaling exponent agrees with the KOC value $1/3$. While the scaling exponent $\zeta_\theta(1) = 0.19$ for the first-order structure function agrees with the BO59 value $1/5$. At first glance, these results seem to be contradicting and confusing. To understand this, we note that in turbulent RBC buoyant forces are exerted on the fluid mainly via thermal plumes.

As revealed by several visualizations, thermal plumes consist of a front with sharp temperature gradient and hence these thermal structures would induce intense temperature increments, which correspond to the pdf tails (Xi et al., 2004; Zhou et al., 2007). It is not surprising that $p^\theta_{\max}(\tau)$ investigated here could not capture efficiently the information of thermal plumes and thus may preclude buoyancy effects (Huang et al., 2011b). We also note that the location of the $p^\theta_{\max}(\tau)$ for a given $p^\theta_\tau(x)$ at a scale τ is around $x \simeq 0$. It thus represents the background fluctuation of the RB convection cell (Huang et al., 2011b).

5.4.4 General remarks on Rayleigh-Bénard convection

As mentioned above, identifying the BO59 scaling is still a challenge. With the help of this probability density function approach, we now understand that, at least in the near side wall region, the temperature background fluctuation is of KOC type, while the thermal plume is responding to the BO59 scaling since this thermal structure possesses a sharp temperature gradient, and thus the buoyancy force is dominant. However, the flow is not fulfilled by the thermal plumes. Therefore, the dynamics at least in the near side wall region can be regarded as a KOC type background fluctuation superposed by thermal plumes, which might obey BO59 type scaling. It is worth pointing out that the overall dynamics are a mixture of the KOC and BO59.

Therefore, knowing how to separate them efficiently is still a challenge. For example, as shown above for the case of Lagrangian turbulence, the Hilbert based method can separate the vortex trapping events efficiently since this structure belongs to the dissipative dynamics and does not overlap with the inertial dynamics. In the RB convection cell, the situation seems more complex: one has to design a special statistical method to solve the above mentioned difficulties. This ideal method should have several abilities: on the one hand, it should remove the large-scale circulation influence/infrared effect; on the other hand, it should separate the KOC background fluctuation and the thermal plumes efficiently.

5.5 Two-dimensional turbulence

5.5.1 Kraichnan's theory of two-dimensional turbulence

Two-dimensional (2D) turbulence is a model for several turbulent flows, such as a possible approximation to the large-scale motion in atmosphere and oceans (Kraichnan and Montgomery, 1980; Tabeling, 2002; Kellay and Goldburg, 2002; Boffetta and Ecke, 2012; Bouchet and Venaille, 2012). The 2D turbulence and its relative problems have attracted a lot of attention in recent years (Falkovich and Lebedev, 1994; Irion, 1999; Chen et al., 2003, 2006; Alexakis and Doering, 2006; Xia et al., 2008, 2011; Boffetta and Musacchio, 2010; Celani et al., 2010; Merrifield et al., 2010; Tran et al., 2010; Kelley and Ouellette, 2011; Khurana and Ouellette, 2012).

Several review papers have been devoted in detail to this topic, for example, Tabeling (2002); Kellay and Goldburg (2002); Van Heijst and Clercx (2009); Boffetta and Ecke (2012); Bouchet and Venaille (2012), to name quote a few. Specifically for the small scale motions, it is believed that there exists a dual cascade, i.e., a forward enstrophy cascade, in which the enstrophy is transferred from large to small scales, and an inverse energy cascade, in which the energy is transferred from small to large scales (Kraichnan, 1967). Two power-law behaviors are thus expected to describe this dual cascade, i.e.,

$$E_u(k) = \begin{cases} C\,(\epsilon_\alpha)^{2/3}\,k^{-5/3}, & \text{when } k_\alpha \ll k \ll k_f \text{ inverse energy cascade} \\ C'\,(\eta_\nu)^{2/3}\,k^{-3}, & \text{when } k_f \ll k \ll k_\nu \text{ forward enstrophy cascade} \end{cases} \tag{5.18}$$

in which $E_u(k)$ is Fourier power spectrum of the velocity, ϵ_α is the energy dissipation by the Ekman friction, η_ν is the enstrophy dissipation by the viscosity, k_f is the forcing scale, in which the energy and enstrophy are injected into the system, k_α

5.5 Two-dimensional turbulence

is the characteristic friction scale, and k_v is the viscosity scale. One can relate the vorticity statistics with the velocity ones by using $E_\omega(k) \sim k^2 E_u(k)$. Therefore, a dual power-law behavior is also expected for the vorticity field, i.e.,

$$E_\omega(k) \sim \begin{cases} k^{1/3}, & \text{when } k_\alpha \ll k \ll k_f \text{ inverse energy cascade} \\ k^{-1}, & \text{when } k_f \ll k \ll k_v \text{ forward enstrophy cascade} \end{cases} \quad (5.19)$$

It is found experimentally that the pdf of the velocity increment $\Delta_\ell u$ is Gaussian when the separation scale ℓ lies in the inverse cascade, indicating nonintermittent behavior on these scales (Kellay and Goldburg, 2002; Tabeling, 2002; Van Heijst and Clercx, 2009; Boffetta and Ecke, 2012; Bouchet and Venaille, 2012).

Note that the classical structure function analysis fails when the slope of the Fourier power spectrum is $\beta \geq 3$ (Frisch, 1995; Huang et al., 2010). This unfortunately is the case of the forward enstrophy cascade in the 2D turbulence (Biferale et al., 2003; Boffetta and Ecke, 2012). Therefore, the intermittent property of the forward energy cascade cannot be verified directly by using the SF analysis (Boffetta et al., 2002; Biferale et al., 2003). Kellay et al. (1998) performed an experimental measurement of the velocity and vorticity field of the 2D soap turbulence. They found that the velocity shows a -3 power-law for the forward enstrophy cascade.

However, the corresponding enstrophy cascade possesses a -2 power-law, which is in contradiction with the theoretical prediction (see Equation [5.20]). Paret et al. (1999) also reported an experiment -3 scaling law for the forward energy cascade. Moreover, they observed for the vorticity increment $\Delta_\ell \omega$ in the forward enstrophy cascade that there is no significant deviation from the Gaussian distribution, i.e., a nonintermittent forward enstrophy cascade. On the contrary, Nam et al. (2000) stated that if an Ekman friction coefficient α is presented, the forward enstrophy cascade is then intermittent (Bernard, 2000). Boffetta et al. (2002) argued that if a passive scalar θ is governed by the same equation as the vorticity and if it is also advected by the same velocity field, it then can be taken as a surrogate of the vorticity ω for the small-scale statistics. They found that the passive scalar θ is intermittent. Moreover, they found that the fitting scaling exponent for the forward enstrophy cascade is strongly dependent on the Ekman viscosity α (see governing Equation [5.20]).

Later, Tsang et al. (2005) studied the intermittency of the forward enstrophy cascade regime with a linear drag. The relative scaling exponent ($\zeta(2q)/\zeta(2)$) provided by the vorticity SF confirms that the forward enstrophy cascade is indeed intermittent for the considered statistical order $0 \leq q \leq 2$ (Tsang et al., 2005). Note that the classical structure function approach is employed in their studies. Biferale et al. (2003) proposed an inverse velocity statistics and applied these in 2D turbulence. They found that the velocity fluctuation cannot be simply described by one single exponent, indicating an intermittent forward energy cascade. Boffetta (2007) reports that the fitting scaling exponent for the forward enstrophy cascade is also strongly dependent on the viscosity ν (see Equation (5.20) below).

More recently, Falkovich and Lebedev (2011) derived analytically the probability density function (pdf) for strong vorticity fluctuations (resp. the tail of the pdf) in the forward cascade. They found that the coarse-grained vorticity $\bar{\omega}$ has a universal asymptotic exponential tail and is thus self-similar without intermittency at least for high-order statistics.

Generally speaking, Kraichnan's theory of 2D turbulence has been partially confirmed by experiments and numerical simulations for the velocity field (Falkovich and Sreenivasan, 2006). However, as mentioned above, the statistics of the vorticity field seem to fail to satisfy the theoretical predictions.

5.5.2 Two-dimensional vorticity field

The 2D Ekman-Navier-Stokes equation is written in terms of a single scalar vorticity field $\omega = \nabla \times \mathbf{u}$ as, i.e.,

$$\partial_t \omega + \mathbf{u} \cdot \nabla \omega = \nu \nabla^2 \omega - \alpha \omega + f_\omega \qquad (5.20)$$

in which ν is the fluid viscosity, α is the Ekman friction coefficient and f_ω is an external source of energy acting on the large scales (Boffetta et al., 2002; Boffetta, 2007). Numerical integration of Equation (5.20) is then performed by a pseudo-spectral, fully dealiased on a doubly periodic square domain of side $L = 2\pi$ at resolution $N^2 = 8192^2$ grid points (Boffetta, 2007). The main parameters are $\nu = 2 \times 10^{-6}$, $\alpha = 0.025$ and $k_f = 100$, in which the energy is injected into the system. The velocity field $\mathbf{u} = \nabla \times \Phi$ is then obtained by solving a Poisson problem $\nabla^2 \Phi = -\omega$, in which Φ is a stream function. Figure 5.56 shows a snapshot of the vorticity field $\omega(x, y)$ on the range $0 \le x, y, \le \pi/2$ (left) and an enlargement part on the range $0 \le x, y, \le \pi/4$ (right). High intensity events are discretely distributed in physical space with a typical wavenumber $k \simeq k_f = 100$, corresponding to 80 grid points. More detail of this database can be found in Boffetta (2007).

5.5.3 Fourier power spectrum and second-order structure function

Figure 5.57a shows the measured Fourier power spectrum $E_\omega(k)$, in which the forcing scale $k_f = 100$ is illustrated by a vertical solid line. Power-law behavior is observed on the range $100 \le k \le 1000$, i.e., $E_\omega(k) \sim k^{-\beta}$, with a scaling exponent $\beta = 1.98 \pm 0.02$ (Tan et al., 2014). The measured β is consistent with the one reported by Kellay et al. (1998). This scaling range is recognized as corresponding to the forward enstrophy cascade. The observed scaling range corresponds to a

5.5 Two-dimensional turbulence

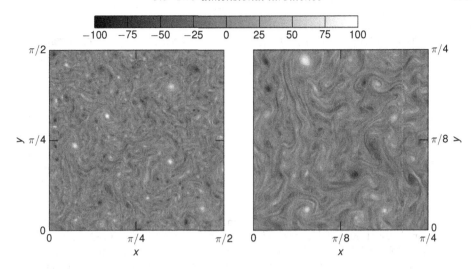

Figure 5.56 A snapshot of the vorticity field $\omega(x,y)$ obtained using a very high resolution direct numerical simulation on the range $0 \le x, y \le \pi/2$ (left) and $0 \le x, y \le \pi/4$ (right). High intensity vorticity events are discretely distributed in space with a typical wavenumber $k \simeq k_f = 100$, corresponding to 80 grid points, approximately.

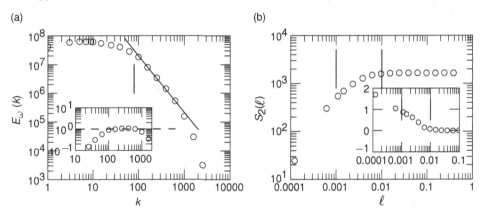

Figure 5.57 a) Measured Fourier power spectrum $E_\omega(k)$, in which the forcing scale $k_f = 100$ is illustrated by a vertical solid line. The inset shows a compensated curve using a scaling exponent -2. b) The corresponding second-order structure function $S_2(\ell)$, in which the power-law range is indicated by a vertical solid line. The inset shows the local slope $\zeta_\omega(2, \ell)$.

spatial scale range $0.001 \le \ell \le 0.01$. We therefore expect a power-law behavior for the second-order structure function,

$$S_2(\ell) = \langle |\Delta_\ell \omega|^2 \rangle \sim \ell^{\beta-1} \tag{5.21}$$

in which $\Delta_\ell \omega = \omega(x+\ell) - \omega(x)$ is the vorticity increment, and β is the scaling exponent from $E_\omega(k) \sim k^{-\beta}$ (Tan et al., 2014). Figure 5.57b shows the measured

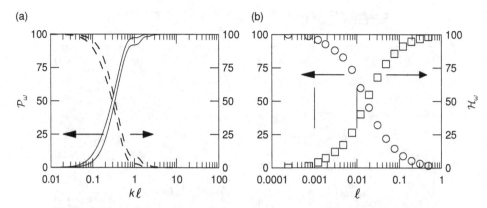

Figure 5.58 a) Measured cumulative function $\mathcal{P}_\omega(\ell, k)$ and the corresponding co-cumulative function $\mathcal{H}_\omega(\ell, k)$ with two separation scale $\ell = 0.001$ (thin solid line) and $\ell = 0.002$ (thick sold line) lying in the forward enstrophy cascade. b) Measured $\mathcal{P}_1(\ell)$ (○) and $\mathcal{H}_1(\ell)$ (□). The forward enstrophy range is indicated by a vertical line.

second-order structure function $S_2(\ell)$, in which the expected power-law range is illustrated by a vertical solid line. Note that there is no clear power-law behavior found for the measured $S_2(\ell)$ for the forward enstrophy cascade on the range $0.001 \le \ell \le 0.01$. To emphasize this point, the local slope of $S_2(\ell)$, i.e., $\zeta_\omega(2, \ell) = d\log_{10} S_2(\ell)/d\log_{10} \ell$, is shown in the inset. There is no plateau observed on the expected power law range, showing the failure of the structure function approach in capturing the scale invariance of this special multiscale field.

To understand more about this observation, we calculate the cumulative function $\mathcal{P}_\omega(\ell, k)$ and co-cumulative function $\mathcal{H}_\omega(\ell, k)$ using the estimated Fourier power spectrum $E_\omega(k)$. Figure 5.58a shows the measured $\mathcal{P}_\omega(\ell, k)$ (solid line) and $\mathcal{H}_\omega(\ell, k)$ (dashed line) for two separation scale $\ell = 0.001$ (thin line) and $\ell = 0.002$ (thick line) lying in the forward enstrophy cascade. Visually, more than 70% of the contribution comes from the large-scale part, for example, $k\ell < 1$. This is because the forward enstrophy cascade is just below the forcing scale $k_f = 100$.

Meanwhile, the observed high intensity vorticity events are observed also with a typical scale $k \simeq 100$. Figure 5.58b shows the measured $\mathcal{P}_1(\ell)$ (○) and $\mathcal{H}_1(\ell)$ (□), in which the scale range of the enstrophy cascade is illustrated by a vertical solid line. Graphically, the measured second-order structure function is dominated by the large-scale part, especially scales close to the forcing scale.

5.5.4 High-order structure functions and extended-self-similarity

Figure 5.59a shows the extended-self-similarity plot of the measured high-order structure functions $S_q(\ell)$ versus $S_2(\ell)$ on the forward enstrophy cascade range

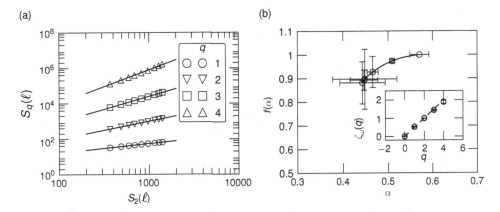

Figure 5.59 a) Extended-self-similarity plot of the high-order structure functions $S_q(\ell)$ versus $S_2(\ell)$ in the forward enstrophy cascade range $0.001 < \ell < 0.01$. b) Measured singularity spectrum $f(\alpha)$ versus α. The inset shows the measured scaling exponent $\zeta_\omega(q)$, in which the dashed line indicates $q/2$.

$0.001 < \ell < 0.01$. Power-law behavior is evident with the help of this technique. The scaling exponent $\zeta_\omega(q)$ is calculated. To emphasize the possibility of the multifractality, the corresponding singularity spectrum $f(\alpha)$ is then estimated. Figure 5.59b shows the measured $f(\alpha)$ versus α, in which the inset shows the measured $\zeta_\omega(q)$. It suggests that the forward enstrophy cascade is weakly intermittent if the multifractality exists.

5.5.5 Hilbert-Huang transform

Empirical mode decomposition

The empirical mode decomposition approach is applied to the vorticity field $\omega(x, y)$ along the x-direction. The extracted number of the intrinsic mode function is found to be $N \in [7, 11]$. Figure 5.60a shows the measured probability density function $p(N)$, in which the Gaussian distribution is illustrated by a solid line. Visually, the measured $p(N)$ agrees very well with the Gaussian distribution. Figure 5.60b shows the measured mean wave number $\bar{k}(n)$. The errorbar is the standard deviation obtained from different realizations. Exponential-law, for example, $\bar{k}(n) \sim \gamma^{-n}$, is observed on the range $1 \le n \le 8$ with a scaling exponent $\gamma = 1.98 \pm 0.02$. This scaling exponent is close to the dyadic value 2.

High-Order Hilbert moments

After retrieving the intrinsic mode functions, the Hilbert spectral analysis is applied to each mode. The qth-order Hilbert moment $\mathcal{L}_q(k)$ is then calculated. Figure 5.61a

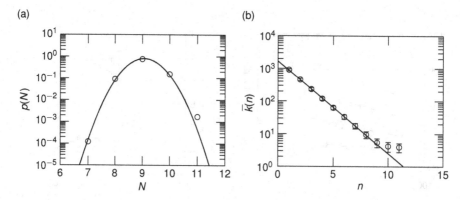

Figure 5.60 a) Experimental probability density function $p(N)$ of the number of extracted intrinsic mode functions, in which the Gaussian distribution is illustrated by a solid line. b) Measured mean wave number $\bar{k}(n)$ for the first 11 modes. Exponential-law is observed with a scaling exponent $\gamma = 1.98 \pm 0.02$, indicating a dyadic filter bank of the empirical mode decomposition.

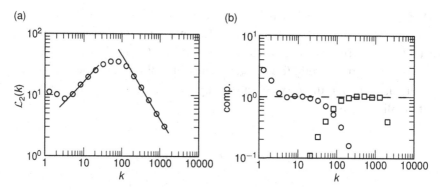

Figure 5.61 a) Measured second-order Hilbert moment $\mathcal{L}_2(k)$. A dual-power-law behavior is observed on the range $3 < k < 20$ for the inverse energy cascade, and $200 < k < 2000$ for the forward enstrophy cascade. b) The corresponding compensated curve using the fitted scaling exponent to emphasize the scaling behavior.

shows the measured Hilbert moment $\mathcal{L}_2(k)$. A dual-power-law is observed on the range $3 < k < 20$ for the inverse energy cascade, and $200 < k < 2000$ for the forward enstrophy cascade (Tan et al., 2014). To emphasize these dual-cascade, the compensated curve using the fitted exponent is shown in Figure 5.61b. A clear plateau is observed. Figure 5.62a shows the high-order moment $\mathcal{L}_q(k)$ on the range $q = 1$ to 4. A dual-power-law behavior is observed for all q considered here. The scaling exponent $\zeta_\omega(q)$ is then retrieved. Figure 5.62b shows the measured scaling exponent $\zeta_\omega^F(q)$ for the forward enstrophy cascade, and $-\zeta_\omega^I(q)$ for the inverse

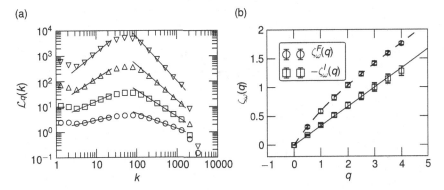

Figure 5.62 a) Measured high-order Hilbert moment $\mathcal{L}_q(k)$ for q between 1 and 4 from bottom to top. b) The corresponding measured scaling exponent $\zeta_\omega^F(q)$ (○) for the forward enstrophy cascade and $-\zeta_\omega^I(q)$ (□) for the inverse energy cascade. For comparison, the solid line is $q/3$ and the dashed line is for a log-Poisson fitting $\zeta_\omega^F(q) = q/3 + 0.45\,(1 - 0.43^q)$.

energy cascade. For comparison, the solid line is $q/3$, and the dashed line is for a log-Poisson-like formula $\zeta_\omega^F(q) = q/3 + 0.45\,(1 - 0.43^q)$ (Tan et al., 2014).

Figure 5.63 shows the Hilbert-based singularity spectrum $f(\alpha)$ for the forward enstrophy cascade (○), and the inverse energy cascade (□). For comparison, the log-Poisson based $f(\alpha)$ is shown as a solid line, and the structure function based is illustrated as △. Visually, the singularity spectrum $f(\alpha)$ of the Hilbert-based forward cascade has a wide range of $0.33 \leq \alpha \leq 0.66$ and $0.57 \leq f(\alpha) \leq 1$, implying an intermittent forward cascade (Tan et al., 2014), while one of the inverse cascade is $0.27 \leq \alpha \leq 0.35$ and $0.80 \leq f(\alpha) \leq 1$, and the structure function based one for the forward cascade is $0.45 \leq \alpha \leq 0.57$ and $0.90 \leq f(\alpha) \leq 1$, respectively. Together with the scaling exponent $\zeta_\omega(q)$, we could draw the conclusion that the inverse cascade is nonintermittent as predicted by the theory. Moreover, the structure-function cannot detect the right multifractality since it is strongly biased by the forcing scale.

5.5.6 General remarks on two-dimensional turbulence

The dual-cascade behavior of the two-dimensional turbulence is a special case in turbulent flows. This dual-cascade is split by the forcing scale, in which the energy and enstrophy are injected into the system. Therefore, the energy balance results in an inverse cascade, while the enstrophy balance leads to a forward cascade. It is also recognized as a result of the dimension reduction. Classical approaches, such as structure function analysis, etc., suffer the scale mixture problem, and are thus made strongly biased by the forcing scale. With the help of the Hilbert

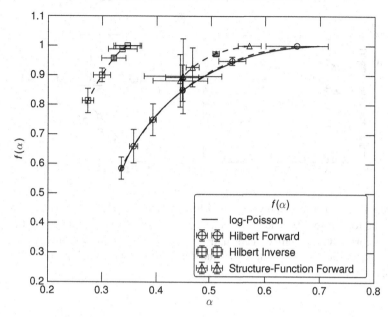

Figure 5.63 Hilbert-based singularity spectrum $f(\alpha)$ for the forward enstrophy cascade (○), and the inverse energy cascade (□). For comparison, the log-Poisson based one is illustrated as a solid line, and the structure function based one for the forward cascade is illustrated by △.

method, the dual-cascade is identified. The multifractality is then recognized by this approach.

5.6 General remarks on scaling behavior in turbulent flows

In this Chapter, experimental and numerical data collected from several turbulent flows have been analyzed as typical multifratal processes. These data are, respectively, Eulerian velocity obtained in a high Reynolds number wind tunnel experiment, temperature data as passive scalar, Lagrangian velocity from high resolution numerical simulation, temperature collected from Rayleigh-Bénard convection cell as active scalar, and vorticity field obtained from a very high resolution numerical simulation. They are different in the sense that they might possess different typical structures. For example, traditional methods, such as structure function analysis, detrended fluctuation analysis, etc., are strongly influenced by energetic structures, no matter whether they are small-scale ones (corresponding to the ultraviolet effect), or large-scale ones (infrared effect) (Huang et al., 2010, 2011a, 2013).

The energetic structures might be different in different flows. For example, in the passive scalar turbulence, it is recognized as ramp-cliff structures; in the Lagrangian

5.6 General remarks on scaling behavior in turbulent flows 161

turbulence, it is the vortex trapping events; in the Rayleigh-Bénard convection, it is the thermal plumes and the large-scale circulation, while in the two-dimensional turbulence, it is the high intensity vorticity structures with the same scale as the forcing scale. Different approaches thus have different performances.

Our experience is that one should check the cumulative and co-cumulative functions to see the influence of energetic structures. If there is no energetic structure, the classical structure function analysis should work. Otherwise, other methods should be tested, such as the Hilbert-based method.

6

Applications: case studies in ocean and atmospheric sciences

In this chapter, we shall consider several case studies from the ocean and atmospheric sciences. Geophysical fluid dynamics is characterized by important variations over a large range of scales. This is true also of ocean sciences, which are often taken as typical example of complex systems, with a huge number of constitutive elements (fluid particles and their scalar properties) interacting through physical, chemical, and biological relations. In the ocean there has been historically one of the first field validation of Kolmogorov's theory, using data from a channel with strong currents between islands (Grant et al., 1962). Since then, other studies have considered oceanic passive scalars, as well as the fluorescence fluctuations as proxy of phytoplankton concentrations. These have compared fluorescence and temperature using Fourier spectra and correlations (Platt and Denman, 1975; Denman, 1976; Denman et al., 1977) and using multifractal approaches (Seuront et al., 1996a,b, 1999). Multifractal studies in the field of oceanology have also been done also using satellite images (Tessier et al., 1993; Lovejoy et al., 2001a,b; Nieves et al., 2007; Pottier et al., 2008; Turiel et al., 2009; de Karman et al., 2011; Renosh et al., 2015).

Superposed to the stochastic variability related to turbulence, coastal marine fields are also subject to deterministic periodic forcing, at fixed scales, such as tidal, diurnal, and annual forcing scales. Hence, time series in the marine sciences can be quite complex and their analysis nontrivial. In the present chapter, we focus on several coastal time series. We first present a coastal marine turbulence time series recorded at 1 Hz in a megatidal sea. We then present a long-term series of water level dynamics recorded every hour, and a database of water quality automatic monitoring recorded at fixed point every 20 minutes, with emphasis on three quantities: temperature, dissolved oxygen, and fluorescence. We finally consider atmospheric examples: atmospheric wind velocity and wind power time series. In each case we perform power spectra, identify some scaling ranges and forcing, and apply several methods among the methods described in the methodological chapters.

6.1 Coastal marine turbulence

We first consider coastal marine turbulence, the interface between land and sea. Close to the shore, waves break and provide a surplus of energy to the marine system. The zone where the bottom has influence on waves is the surf zone (Svendsen, 1987; Battjes, 1988; Schmitt et al., 2009). It is a zone where tidal waves input energy into the turbulence system. The swash zone is closer to the shore, where waves splash and dissipate into the beach. Hence the surf zone is a zone of strong erosion and energy, sediment and biological transport, and resuspension processes. Here we will consider the scaling properties for scales between the tidal scale (around 6 hours) and the breaking of waves (around 10 seconds). For these scales, turbulence is forced by the tidal energy and influenced by the bottom.

6.1.1 Presentation of the data

The data were recorded in the Eastern English Channel, near Boulogne-sur-mer, France, at location ($50°45.676N$, $01°35.117E$) from 25 to 28 June 2012, using an Acoustic Doppler Velocimeter (ADV) from Nortek (Nortek Vector) at 1 Hz, together with other instruments fixed on a platform moored on the sea bed (see Renosh et al. [2014] for a presentation of the data set). The English Channel is a mega-tidal sea with a tidal range at this location of 3 to $9\,m$, with tidal currents up to $1.0\,m/s$ (Desprez, 2000; Seuront and Schmitt, 2005; Korotenko et al., 2012).

The ADV measurements provide us with the 3D velocity vector averaged on a small volume of about 250 mm^3 at a distance of 5 cm from the ADV probe, with an accuracy of 0.5% of the measured value, and the water height in meters, proportional to the local pressure. A one-day portion of the velocity and height datasets is shown in Figure 6.1, emphasizing two tidal cycles.

6.1.2 Power spectra

Power spectra have been estimated for both velocity components (Figure 6.2). We consider here the scaling property of the U and V components between $12s$ and 1 hour, shown in the figure with the -0.58 power-law scaling. This exponent (already found in Renosh et al. [2014]) is far from the 5/3 exponent for fully developed turbulence. It could be related to the -1 spectrum which has been theoretically predicted close to the wall, by several researchers (Panchev, 1972; Kader and Yaglom, 1984; Perry et al., 1986; Katul et al., 1995; Katul and Chu, 1998). Since U and V components have the same spectrum, we will consider below only one time series (U). On the same figure, for high frequencies (from 3 to 10 s) there is a localized forcing associated to wave-breaking scales (Svendsen, 1987; Schmitt et al., 2009). For lower frequencies, the forcing associated to the tidal cycle and harmonics are visible (12h, 6h).

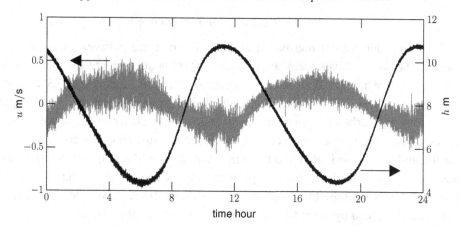

Figure 6.1 A 24 hours portion of the U-velocity time series, superposed on the height information, indicating the tidal range.

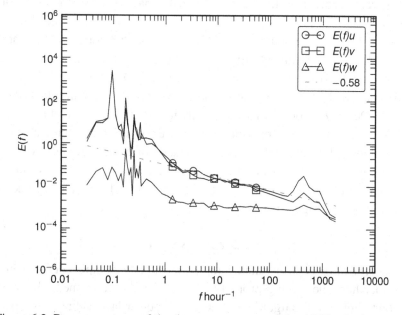

Figure 6.2 Power spectrum of the three components of the ADV velocity, superposed to the water level power spectrum. A power-law fit $f^{-0.58}$ is superposed on U and V power spectra of the velocity.

Due to the energy input at small scales associated to the wave breaking, methods such as the structure function, detrended structure function, and autocorrelation function of the increment cannot be applied to retrieve scaling properties. We have thus considered here only the arbitrary order Hilbert spectral analysis. Figure 6.3 shows the result for moments from 0 to 4. A scaling range is obtained for scales

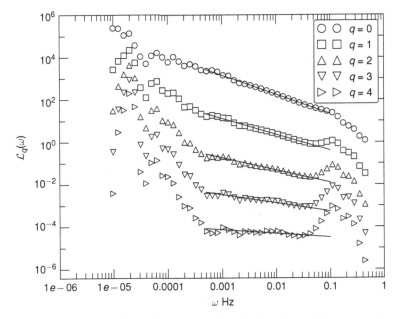

Figure 6.3 Arbitrary order Hilbert spectral analysis of the U-velocity data in the bottom boundary layer: moments of order 0 to 4 are displayed. The scaling exponents for scales from 12 s to 1 hour are estimated.

from 12 s to 1 hour. The scaling is even larger for the moment of order 1, with a slope of -0.74, corresponding to a negative H values: $H = \zeta(1) = \xi(1) - 1 = -0.26$. This recalls results recently published, where negative H values are obtained using the Haar wavelet (Lovejoy and Schertzer, 2012, 2013). This shows that the arbitrary order HSA method can also retrieve negative H values. Such result indicates that small-scale fluctuations are larger than large-scale ones, still in a scale-invariant framework. The generalized moment function $\xi(q)$ is shown in Figure 6.4: it is slightly nonlinear and concave, indicating a signature of multifractality. Such scaling regime is situated between two forcing: a small-scale forcing associated to the wave breaking around a few seconds, and the tidal forcing associated to 6 hours (and its subharmonics at 3 hours). Here the scaling range which is found could be related to the boundary effect leading to -1 scaling, or to another effect in relation to the two forcing.

6.2 Water-level dynamics

Water level is a complex signal, with a strong deterministic part and a stochastic part. The deterministic part corresponds to the astronomical influence in relation to the shape of oceanic basins. The stochastic component corresponds to the influence

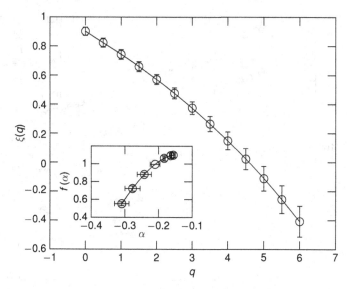

Figure 6.4 The generalized moment function $\xi(q)$ estimated for the U velocity data in the bottom boundary layer, for scales from 12 s to 1 hour. The inset shows the singularity spectrum.

of meteorological phenomenon, and results from nonlinear interactions between pressure, temperature, winds, wind waves, and turbulence (Pugh, 2004; Weisse, 2010). In coastal regions, a better comprehension of the water-level dynamics, at various time scales, is important in being able to understand and model surge events and long-term water-level increases.

6.2.1 Presentation of the data

Here we consider hourly measurements and model outputs, performed at Boulogne-sur-mer (Northern, France) in the Eastern English Channel by SHOM (Service hydrographique et océanographique de la Marine, National hydrographic Service, France), between 7 September 1973 and 1 January 2012 (38% of which are missing and contain 208,971 data points).

The model is computed by considering a Fourier decomposition of the potential generating the tide, where the amplitudes of the first terms in the decomposition depend on the local hydrographic conditions. These terms are kept confidential, and a model time series is generated, which is deterministic and can be used to predict, for a given date, the expected tidal level. This signal has no stochastic component and the true measured signal is the sum of the deterministic part and stochastic terms. The difference between the model and the measurements can reach 2 meters and, if the difference is positive and happens during a high tide, the surge can lead

6.2 Water-level dynamics

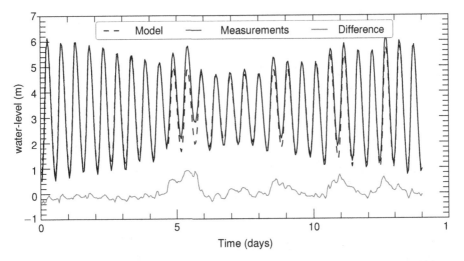

Figure 6.5 A 14 days portion of the modelled water-level time series at Dunkirk, and the measured data (above, superposition of the model and measurements). The time series difference (measurements-model) shows that the superposition is not perfect (bottom).

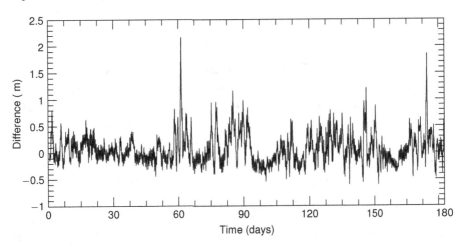

Figure 6.6 A 6 months portion of the difference time series at Dunkirk (measurements-model), showing its stochastic and multiscale dynamics.

to a coastal flood. Figure 6.5 represents a small portion of the data (14 days) as well as the difference time series. Figure 6.6 represents a longer time series (6 months duration) of the difference data, showing its stochastic and multiscale dynamics. Figure 6.7 represents the pdf of the difference data: it shows that it can be modeled using an exponential function in the form $p(x) = A \exp(bx)$, with $A = 0.52$ and $b = 7.7$ for $x < 0$ and $A = 0.37$ and $b = -6.2$ for $x > 0$. The pdf is asymmetric, with a slower decrease for positive values.

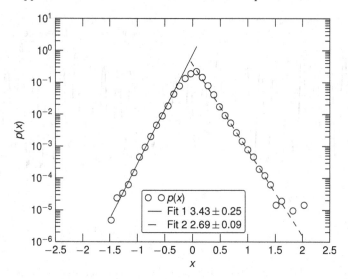

Figure 6.7 The pdf of the difference data, computed over the whole time series, with two exponential fits, of equation $p(x) = A \exp(bx)$, with $A = 0.52$ and $b = 7.7$ for $x < 0$ and $A = 0.37$ and $b = -6.2$ for $x > 0$.

6.2.2 Power spectra

Power spectra are shown in Figure 6.8a superposed for the model and the measurements. Since classical Fourier analysis needs regularly spaced data (for the use of the Fast Fourier Transform, FFT), we have considered here a subsample of the full time series, containing no missing data: it is a 37,613 data points long portion (from 23 April 2003 to 7 August 2007) and hence large enough to perform statistical analysis. The superposition is rather limited: the model spectrum is close to the experimental spectrum only for some harmonics corresponding the the tidal cycle. Several pikes are visible for the model data: these correspond, from low frequency to high frequency, to Solar annual, Solar semiannual, Lunar monthly, Lunisolar fortnightly (around 14.5 days), Solar diurnal, Lunar semidiurnal (M_2), and other harmonics linked with the tide, such as M_4, M_6, and M_8. For lower frequencies, there is a big gap between the energy spectrum of data and model, which has a very low level of energy compared to the data. This is highlighted using the ratio of experimental to model spectra in Fig. 6.8b: the ratio reaches 10^5 (around the scale of 30 days) and between 1.25 and 166 days, it is mostly between 100 and 1000, indicating that between 1 day and 6 months the natural variability is much larger than modeled.

The experimental spectra show power-law behavior over more than one decade, with a slope estimated as 1.45. The origins of such spectra are diverse, as mentioned earlier in this chapter, but many of the causes can be linked to the direct or indirect

6.2 Water-level dynamics

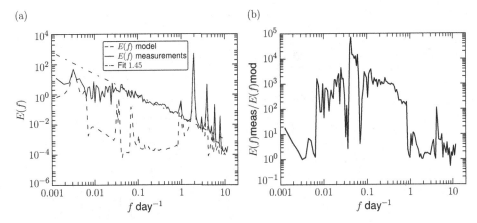

Figure 6.8 Left: Power spectra of model and measurements of water-level data (from 23 April 2003 to 7 August 2007). The superposition is good only for high-frequency cycles. For scales larger than one day the superposition is not good, and the model does not reproduce the dynamics revealed by the data. Right: This is highlighted by the ratio of spectra, which is mostly between 100 and 1,000, indicating that between one day and six months the natural variability is much larger than modeled.

influence of turbulence (winds, pressure, temperature, etc.), explaining such power-law behavior and spectral slope.

6.2.3 Intermittency analysis

An intermittency analysis has been performed for the water-level measurement time series for scales from one to 25 days, using the EMD-HSA method. As for the *in situ* velocity time series, the high frequency energetic forcing found here prevents the use of many scaling methods. The scaling range for moments from $q = 0$ to 4 is shown in Figure 6.9. It is found that the scaling is good for low-order moments: for a moment of order 1, there is a scaling law of slope 1.23, corresponding to a value $H = 0.23$, for scales from 1 to 40 days. For $q = 3$ and 4, the scaling is valid over a reduced range of scales. To estimate scaling exponents over the same range, a range from 30 hours to 28 days has been chosen. This gives the estimate $\xi(q)$; the function $\zeta(q) = \xi(q) - 1$ is shown in Figure 6.10, together with the associated singularity spectrum. It is clearly non-linear, and decreases for moments larger than 2.

Such dynamics characterizes the stochastic part, influenced by meteorology, of the water-level signal. The scaling property of such a signal can be related in some way to turbulence, through atmospheric pressure, temperature, and velocity, and both fields can have multiscale fluctuations through interactions with active or passive scalar turbulence. Here, due to the very intense energy associated with the

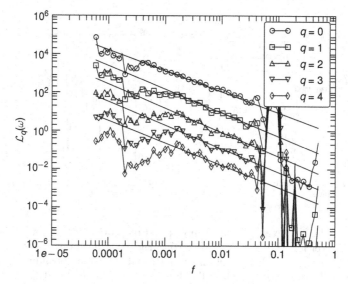

Figure 6.9 Scaling of the moments using Hilbert spectral analysis of the water-level time series.

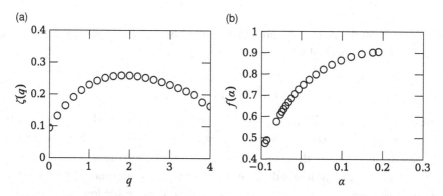

Figure 6.10 a) The moment function $\xi(q)$ estimated for the stochastic part of water-level fluctuations, for scales between 1 and 25 days. b) the associated singularity spectrum.

tidal frequency, visible in Figure 6.8, extraction of the scaling properties of the stochastic part of the signal is not a clear task, and it is a positively interesting feature of the Hilbert spectral analysis to be able to extract such information.

6.3 Water quality automatic monitoring

We consider here automatic sampling done using fixed-point systems in coastal waters, measuring various biogeochemical parameters. These systems are usually in position for several years, enabling us to monitor the dynamics of these parameters

6.3 Water quality automatic monitoring 171

over a large range of time scales (Dickey, 1991; Chavez, 1997; Chang and Dickey, 2001; Nam et al., 2005; Dur et al., 2007; Schmitt et al., 2008; Zongo and Schmitt, 2011; Zongo et al., 2011).

6.3.1 Presentation of the data

Here we consider a dataset belonging to the MAREL network (Mesures automatisées en réseau pour l'environnement littoral - network of automatic measurements for the littoral environment; Blain et al., 2004; Dur et al., 2007; Schmitt et al., 2008; Zongo and Schmitt, 2011; Zongo et al., 2011). We consider the MAREL Carnot automatic system from this network, installed at a fixed position ($50°44.42N$, $01°34.06E$) in the Eastern English Channel near Boulogne-sur-mer (France) and operated by IFERMER.

About fifteen biogeochemical parameters are recorded every twenty minutes; we shall consider here only four of them: air temperature, water temperature, fluorescence, and dissolved oxygen (percentage of saturation). The latter three are important quantities for ecosystem studies and physics-biology couplings, and their multiscale fluctuations may be compared to a passive scalar such as temperature. We will also consider the temperature difference between the air and water. The data considered here is from a three years portion, recorded from 1 January 2007 to 31 December 2009. Theoretically, the total is 78,911 data, but there are various missing due to failure of the measuring devices or maintenance activities. Overall, the percentage of the present values (Figure 6.11) are 81.6% for dissolved oxygen, 88.3% for fluorescence, 88.1% for air temperature and 90.0% for water temperature. In Figure 6.11a, the temperature difference is also displayed. All quantities, except the dissolved oxygen time series, show a clear annual cycle. All series have strong fluctuations at many different scales. The fluorescence time series shows an annual spring bloom associated with *Phaeocystis globosa* micro algae dynamics.

6.3.2 Hilbert spectral analysis

Since there are missing data, Fourier analysis has not been performed for these series. We shall first perform Hilbert spectral analysis on the temperature series. Figure 6.12a shows the HSA moment of order $q = 1$, emphasizing a value of $H = 0.2$ and 0.4 for air and water temperature, respectively. The scaling range is from 6 hours to 1 month for the air temperature, and from 1 day to 3 months for the water temperature. Both spectra join for scales larger than three months; the daily cycle is visible for both series; for the water temperature the tidal cycle is also visible. For scales below 1 day the air temperature spectrum becomes steeper than the water spectrum. The difference spectrum is very close to the air spectrum,

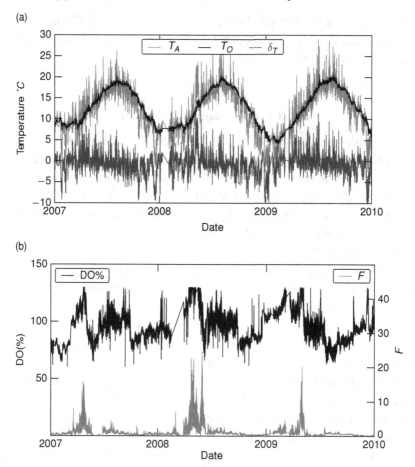

Figure 6.11 Presentation of the 4 MAREL time series considered here. a) Time series of air and water temperature. There is an annual cycle in both cases. The temperature difference (air-water) is also shown. b) The dissolved oxygen (percentage of saturation) and fluorescence time series. The oxygen does not display a clear annual cycle; the fluorescence has an annual spring bloom.

except for scales larger than three months. We have also displayed in Figure 6.12b the probability density function of the difference series. The maxima are -12 and $+11\ °C$, with a negative mean value ($-0.76°C$). The symmetric is also plotted as a dotted line, showing that the pdf is not symmetric, with negative values (ocean water colder than the atmosphere) more frequent than positive values of the same magnitude. The shape is roughly an exponential.

We shall also consider the dissolved oxygen (DO) dynamics. Dissolved oxygen is an important quantity for ecological studies, since it is essential for aquatic life. As a key parameter, it is influenced by many factors, such as water temperature and salinity; photosynthesis; respiration; turbidity and light penetration; turbulence

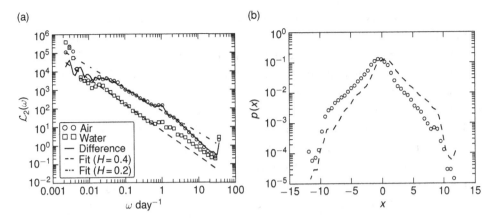

Figure 6.12 a) Hilbert spectral analysis, for the moment $q = 1$ of air, water temperature and the difference. Power fits are also shown, corresponding to $H = 0.2$ and 0.4 for the air and water temperature series, respectively. b) The probability density function of the difference $\delta T = T_A - T_O$ The dotted line is the symmetric. It shows that the pdf is not symmetric and is roughly an exponential.

and air-sea exchanges. As a result, the dynamics of DO are likely to be complex, with variability at many different scales. Conversely, this parameter can be expected to display stochastic variations due to the influence of turbulence in the transport and mixing of DO, as well as the air-sea exchanges. Equally, there could be strong deterministic periodic forcing, with the diurnal cycle of photosynthesis, and the semidiurnal effect of the tide. It is therefore clearly interesting to measure high frequency time series of DO. We study here an oxygen saturation time series, estimated as a percentage of DO relative to the dissolved oxygen at equilibrium, for the same temperature and salinity. We use the oxygen solubility equation given in Garcia and Gordon (1992):

$$SaO_2 = \frac{100DO}{\exp\left(P_1(T) + SP_2(T) + C_0 S^2\right)} \quad (6.1)$$

where SaO_2 is the percentage of saturated oxygen, DO is the dissolved oxygen concentration, and the denominator is the nonlinear fit expressing the oxygen solubility. S is salinity, T is temperature, and P_1 and P_2 are two polynomial developments: P_1 of degree five, and P_2 of degree three. The coefficients providing the best fit are given in Garcia and Gordon (1992). Values exceeding 100 % correspond to the supersaturation associated with mixing or high primary production, while those below 100 % correspond to undersaturation associated with oxygen depletion. When the undersaturation reaches values below 50%, there is a danger to aquatic animal life.

Moreover, we also consider the fluorescence time series, which is a biological parameter, and proxy of phytoplankton concentration, which plays an important role in aquatic ecosystems and carbon dynamics. Phytoplankton is influenced by temperature and needs light for its growth. It has a strong annual cycle, with regular spring blooms (Figure 6.11). It may also, however, be influenced by turbulent transport and nutrients.

Figure 6.13 shows the Hilbert spectral analysis of both DO and F series, for the first moment. A clear scaling relation is visible, with $H = 0.42$ and 0.5 for the dissolved oxygen and fluorescence series, respectively. In both cases the scaling range is two decades wide, for scales larger than 1 day. The daily cycle is visible on the DO data, but not on the fluorescence data, displaying a tidal influence.

For temperature series we have not displayed here the intermittency exponent, as the scaling range is not large enough to accurately estimate larger moments statistics. For the dissolved oxygen and fluorescence, we have performed Hilbert spectral analysis considering moments from 0 to 4. Scaling of the moments from 0 to 4 is shown in Figure 6.14. The scaling range chosen is the following: from 1.8 to 315 days for the dissolved oxygen and from 1.2 to 167 days for the fluorescence. This corresponds to a range of scales between two deterministic forcing: the daily and the annual cycles. The resulting intermittency scaling exponents are shown in Figure 6.15.

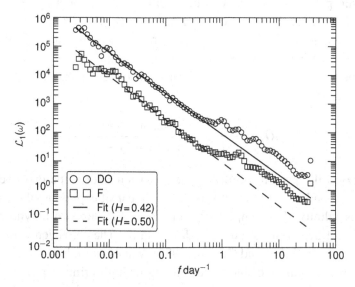

Figure 6.13 Hilbert spectral analysis, for the moment $q = 1$ of dissolved oxygen and fluorescence data. Power fits are also shown, corresponding to $H = 0.42$ and 0.5 for the dissolved oxygen and fluorescence series, respectively.

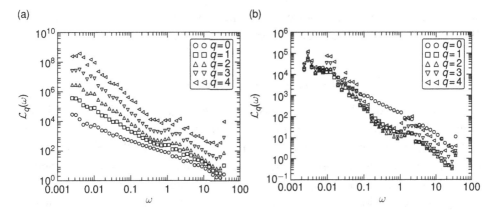

Figure 6.14 Scaling of Hilbert spectral analysis for the dissolved oxygen (left) and fluorescence data (right). The scaling range considered is between the daily and annual cycles.

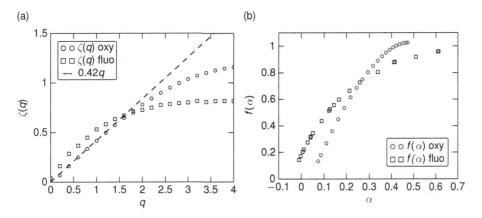

Figure 6.15 Left: Scaling exponents $\zeta(q) = \xi(q) - 1$ of both dissolved oxygen and fluorescence data. There is a much stronger intermittency for the latter series. Right: Associated singularity spectrum.

6.4 Atmospheric wind velocity

We analyze here atmospheric wind time series. In the classic book Monin and Yaglom (1971), many atmospheric power spectra are displayed. Many more are also given in the recent book by Lovejoy and Schertzer (2013). Kolmogorov spectra with 5/3 power-law slope for the wind velocity, and multifractal scaling properties, have been found in many studies (Schmitt et al., 1992, 1994; Katul et al., 1995; Boettcher et al., 2007; Calif and Schmitt, 2012; Morales et al., 2012; Calif and Schmitt, 2014). However, due to the presence of the ground, some measurements in the atmospheric surface layer in neutral conditions, as well as theoretical studies, have found a scaling with a power-law slope closer to $\beta = -1$ (Katul and Chu, 1998; Nikora, 1999; Nickels et al., 2005; Katul et al., 2012).

6.4.1 Dataset and power spectrum

The dataset considered here is a windspeed series measured in the island Guadeloupe in the West Indies, with an ultrasonic anemometer mounted on a tall mast 38 meters from the ground. The sampling frequency is 20 Hz, and the record was taken over during 28 hours in July 2005 (see Calif and Schmitt [2012] for a presentation of the dataset and a more complete analysis using a larger dataset). A total of 2.10^6 data points is considered here: the data range is from 2.57 to 8.37 m/s, with a mean value of 5.67 m/s. A one-hour portion of the dataset is shown in Figure 6.16, illustrating the multiscale large fluctuations of the wind velocity.

The corresponding Fourier power spectrum is displayed in Figure 6.17. A power-law range is found on the range 10^{-4} to 10 Hz with a slope of 1.27 ± 0.04. The compensated spectrum shows precisely that this power-law is found on the range 10^{-4} to 10^{-1} Hz (10 seconds to about 3 hours). Such a slope of $\beta = 1.27$ is intermediary between 1 and 5/3; it seems that there is no theoretical result that could be put forward to explain such a value.

6.4.2 Multiscaling analysis using structure functions and Hilbert spectral analysis

Structure function analysis is performed on this data set, and the scaling range is shown in Figure 6.18: a good scaling is found on the range $5 < \tau < 500$ seconds.

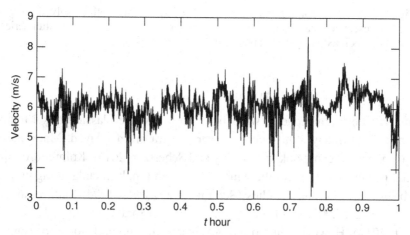

Figure 6.16 A one-hour portion of the atmospheric wind velocity measurements.

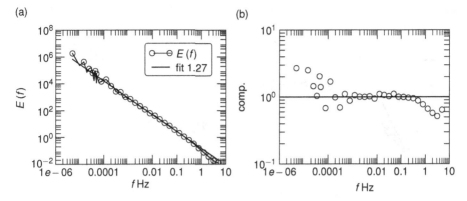

Figure 6.17 a) Power spectrum of the wind velocity measurements, where a power-law range is found on the range 10^{-4} to 10 Hz with a slope of 1.27 ± 0.04. b) The compensated spectrum, showing that the power-law is found precisely for frequencies on the range 10^{-4} to 10^{-1} Hz.

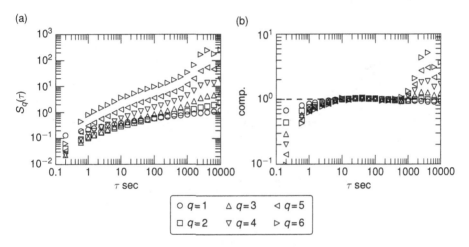

Figure 6.18 a) The structure functions applied on the wind velocity series, for moments from 1 to 6; b) the compensated structure functions, showing the scaling range, from 5 to 500 seconds.

The corresponding scaling exponents are estimated on this range and displayed in Figure 6.19. A strongly nonlinear curve is obtained, indicating a strongly intermittent field. Such a curve is not the same as found in classical homogeneous turbulence, where $\zeta(3)$ is close to 1; here $\zeta(3)$ is smaller than 0.4 and the exponents seem to saturate for larger moments, to a constant value slightly larger than 0.4.

Arbitrary order Hilbert spectral analysis is also performed. Scaling is found for scales from 0.003 to 1 Hz (Figure 6.20), corresponding to a range of time scales from 1 second to 5 minutes. The scaling exponents $\xi(q)$ are estimated on this

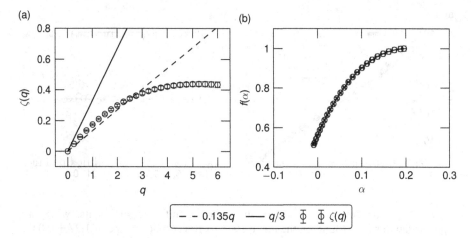

Figure 6.19 a) The scaling exponent $\zeta(q)$ estimated from the wind velocity time series using structure functions, for scales from 5 to 500 seconds; b) the associated singularity spectrum.

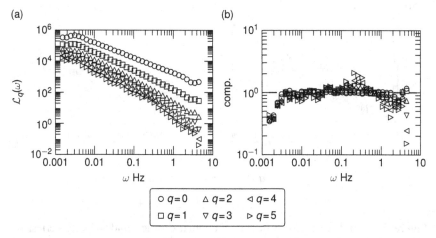

Figure 6.20 a) Arbitrary order Hilbert spectral analysis applied on the wind velocity data, for moments from 0 to 5; the scaling range is confirmed by the compensated values in b).

frequency range; the resulting $\zeta(q) = \xi(q) - 1$ function is shown in Figure 6.21: it has a shape similar to the one found using the structure functions.

The Hurst scaling exponent $H = \zeta(1)$ is here close to 0.2; there is not dimensional analysis nor theoretical cascade model proposed (to our knowledge) that could explain such a value. It could be a boundary layer effect. However, such a low H value is associated with a strong intermittency over a large range of time scales.

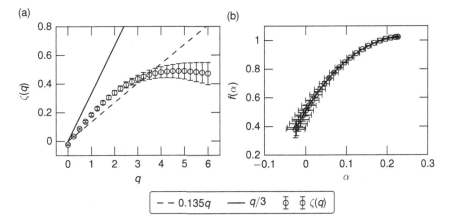

Figure 6.21 a) The scaling exponents $\zeta(q) = \xi(q) - 1$, obtained from the arbitrary order Hilbert spectral analysis of wind velocity, with a fit over the range 0.003 to 1 Hz; b) the associated singularity spectrum.

6.5 Wind power time series

Wind energy is a fast growing global technology (Gipe, 1995). It is a clean energy and renewable resource, however, one of the main problems associated with wind energy is its high volatility, also called in this domain "intermittency." The meaning of the word *intermittency* in this community is not the same as in the homogeneous turbulence community: in the wind energy field, and more generally associated to renewable energies (marine and solar), this corresponds to an idea of small-scale nonstationarity. One of the main issues in this respect is to be able to predict wind energy production, at least on small scales (Landberg, 1999; Sanchez, 2006; Costa et al., 2008). In fact the main reason for wind power nonstationarity is the turbulent character of the atmospheric wind velocity (Peinke et al., 2004). Indeed, such systems require large mean wind velocity; in such cases the Reynolds number is also very large and as a consequence the wind fluctuations are also large, with intermittent bursts.

The transfer from wind velocity to power production is a nonlinear one; the average transfer function, shown in Figure 6.22, is a conditional average. Instantaneous values are scattered around this mean curve (Sanchez, 2006; Gottschall and Peinke, 2008). It has four zones: below a given threshold (here for illustration purposes, 3 m/s), there is no production since the velocity is too low and the wind turbine is not operating. For the increasing part, the power is proportional to the cube of the velocity; then around a given value (here for illustration, it is 10 m/s) there is a plateau since the wind turbine reaches a maximum rotation velocity. For excessive values of the velocity, for safety reasons the windmill is not operating and the production goes to zero again.

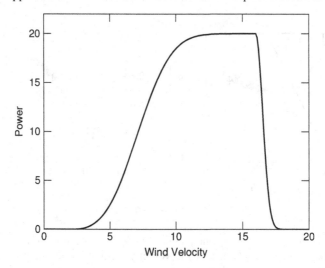

Figure 6.22 A typical transfer function from the wind velocity to the wind power, with arbitrary numerical values given here as illustration. It should be seen as a conditional average – instantaneous values do not follow this curve. There are four zones: below a given threshold (here 3 m/s), there is no production since the velocity is too low. For the increasing part, the power is proportional to the cube of the velocity; then around a given value (here 10 m/s) there is a plateau and for excessive values of the velocity (here 16 m/s), for safety reasons the windmill is not operating and the production goes to zero again.

Power fluctuations can be can be simulated using wind models (Sørensen et al., 2002; Calif and Schmitt, 2012). A question of interest is also to consider if reducing the variability of output power is possible by considering interconnected wind plants (Katzenstein et al., 2010). Here we show an example of wind power data produced by a wind farm, and the multiscaling properties of its power production.

6.5.1 Data and power spectrum

The data come from the same site as the atmospheric data of the previous section: the wind energy production site of Petit-Canal, an island in the French West Indies. The aggregate power output produced by this 10 MW wind farm was recorded with a sampling frequency of 1 Hz, over one year (January 2006 to January 2007). This data set has been used in previous studies (Calif et al., 2013; Calif and Schmitt, 2014).

The data set is shown in Figure 6.23: it shows a large variability with very large fluctuations at very small scales. When the wind input is too low, the wind farm has stopped and is not producing energy. Only the first part of the series is analyzed here. It spans from 21.9 to 1530 kW, with a mean of 585. The corresponding Fourier

6.5 Wind power time series

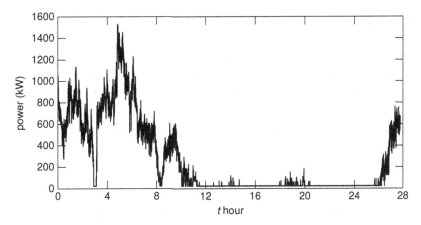

Figure 6.23 The power output time series studied here, showing its high multi-scale variability, and also some periods when the wind farm is stopped, when the wind input is below a given threshold. Only the first part of the time series is studied here.

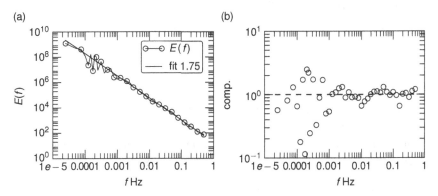

Figure 6.24 a) Power-law scaling of the output power series, with a slope of 1.75± 0.04, close to 5/3; b) compensated spectrum showing the power-law range.

power spectrum is shown in Figure 6.24. It is scaling with a power-law slope of 1.75 ± 0.04, close to 5/3, for frequencies from $5\,10^{-4}$ to 0.5 Hz, from 2 seconds to around 33 minutes. Such a slope for wind power production, close to 5/3, has already been shown in several works (Apt, 2007; Katzenstein et al., 2010; Calif et al., 2013; Calif and Schmitt, 2014).

6.5.2 Scaling analysis using structure functions and Hilbert spectral analysis

Since the power output time-series displays scaling properties with a slope close to K41 turbulence, it is likely that such series also exhibits multiscaling properties.

This was proposed by Calif and Schmitt (2012), transforming wind velocity measurements into modeled power time series using the power transfer curve. It was also checked directly on wind power time-series (Calif et al., 2013; Milan et al., 2013; Calif and Schmitt, 2014). We show here the result on our wind power dataset. The analysis is done using structure functions as well as Hilbert spectral analysis.

Figure 6.25 displays the scaling of structure functions, for moments from 1 to 6. There is a very nice scaling over almost 4 decades, from 1 to 5000 seconds=83 minutes. Scaling exponents estimated over this range are shown in Figure 6.26.

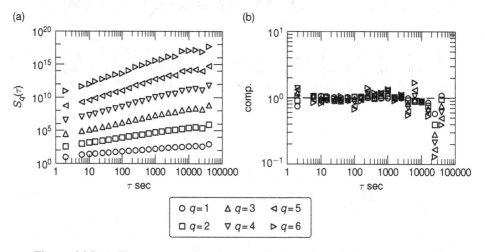

Figure 6.25 a) The structure functions applied on the wind power series, for moments from 1 to 6; b) the compensated structure functions, showing the scaling range, which is quite large, over almost 4 orders of magnitude.

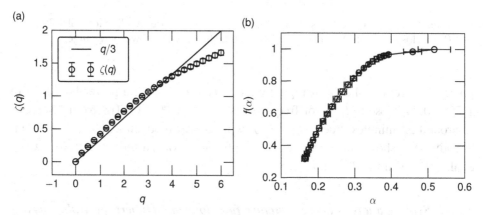

Figure 6.26 a) The scaling exponent $\zeta(q)$ estimated from the wind power time series using structure functions, for scales from 1 to 5,000 seconds. b) The associated singularity spectrum.

6.5 Wind power time series

The nonlinear curve obtained for the $\zeta(q)$ curve is similar to the one found for the velocity field in homogeneous turbulence - the associated singularity spectrum is shown in Figure 6.26b.

Arbitrary order Hilbert spectral analysis is also performed; Figure 6.27 shows the result for moments from 0 to 5, with a scaling found on the range 10^{-4} to 0.2 Hz; Figure 6.28 displays $\zeta(q) = \xi(q) - 1$, which is nonlinear and concave, and quite similar to $\zeta(q)$ obtained using structure functions. For such dataset without deterministic forcing, both methods are giving close results.

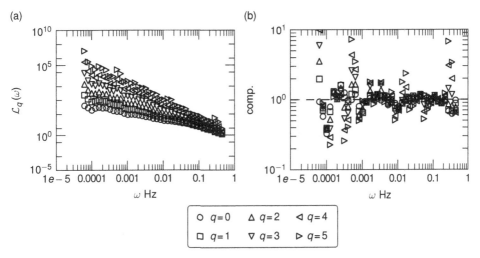

Figure 6.27 a) Arbitrary order Hilbert spectral analysis applied on the wind power data, for moments from 0 to 5. There is a nice scaling, confirmed by the compensated values in b).

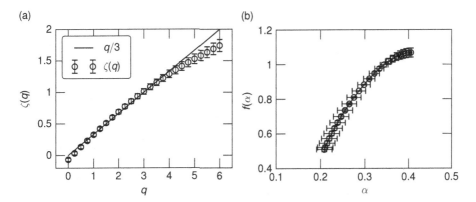

Figure 6.28 a) The scaling exponents $\zeta(q) = \xi(q) - 1$, obtained from the arbitrary order Hilbert spectral analysis, with a fit on the range 10^{-4} to 0.2 Hz. b) The associated singularity spectrum.

Here these results show that the wind power time-series are indeed intermittent, using the meaning of the word from the field of homogeneous turbulence: they have large variability, with correlated fluctuations possessing multifractal scaling. Such results, found rather recently (Calif et al., 2013; Milan et al., 2013; Calif and Schmitt, 2014), indicate that it should be possible to exploit the correlation structure of such multifractal fields for predictability purposes (Marsan et al., 1996; Perpete and Schmitt, 2011), providing short-term predictions using adequate algorithms.

References

Abry, P., Chainais, P., Coutin, L., and Pipiras, V. 2009. Multifractal random walks as fractional Wiener integrals. *IEEE Trans. Inform. Theory*, **55**(8), 3825–3846.

Ahlers, G., Grossmann, S., and Lohse, D. 2009. Heat transfer and large scale dynamics in turbulent Rayleigh-Bénard convection. *Rev. Mod. Phys.*, **81**(2), 503–537.

Alexakis, A., and Doering, C. R. 2006. Energy and enstrophy dissipation in steady state 2D turbulence. *Phys. Lett. A*, **359**(6), 652–657.

Anh, Vo V., Leonenko, Nikolai N., and Shieh, Narn-Rueih. 2008. Multifractality of products of geometric Ornstein-Uhlenbeck-type processes. *Adv. Appl. Prob.*, **40**, 1129–1156.

Anselmet, F., Gagne, Y., Hopfinger, E. J., and Antonia, R. A. 1984. High-order velocity structure functions in turbulent shear flows. *J. Fluid Mech.*, **140**, 63–89.

Applebaum, D. 2004. *Lévy processes and stochastic calculus*. Cambridge University Press.

Apt, J. 2007. The spectrum of power from wind turbines. *J. Power Sources*, **169**(2), 369–374.

Arnéodo, A., Baudet, C., Belin, F., Benzi, R., Castaing, B., Chabaud, B., Chavarria, R., Ciliberto, S., Camussi, R., and Chilla, F. 1996. Structure functions in turbulence, in various flow configurations, at Reynolds number between 30 and 5000, using extended self-similarity. *Europhys. Lett.*, **34**(6), 411–416.

Arnéodo, A., Benzi, R., Berg, J., Biferale, L., Bodenschatz, E., Busse, A., Calzavarini, E., Castaing, B., Cencini, M., Chevillard, L., Fisher, R. T., Grauer, R., Homann, H., Lamb, D., Lanotte, A. S., Lévêque, E., Luthi, B., Mann, J., Mordant, N., Muller, W.-C., Ott, S., Ouellette, N. T., Pinton, J.-F., Pope, S. B., Roux, S. G., Toschi, F., Xu, H., and Yeung, P. K. 2008. Universal intermittent properties of particle trajectories in highly turbulent flows. *Phys. Rev. Lett.*, **100**(25), 254504.

Arrault, J., Arnéodo, A., Davis, A., and Marshak, A. 1997. Wavelet based multifractal analysis of rough surfaces: application to cloud models and satellite data. *Phys. Rev. Lett.*, **79**(1), 75–78.

Ashkenazi, S., and Steinberg, V. 1999. Spectra and statistics of velocity and temperature fluctuations in turbulent convection. *Phys. Rev. Lett.*, **83**(23), 4760–4763.

Bacry, E., and Muzy, J.-F. 2003. Log-infinitely divisible multifractal processes. *Commun. Math. Phys.*, **236**(3), 449–475.

Bacry, E., Delour, J., and Muzy, J. F. 2001. A multifractal random walk. *Phys. Rev. E*, **64**(2), 026103.

Balocchi, R., Menicucci, D., Santarcangelo, E., Sebastiani, L., Gemignani, A., Ghelarducci, B., and Varanini, M. 2004. Deriving the respiratory sinus arrhythmia from the heartbeat time series using empirical mode decomposition. *Chaos Soliton Fract.*, **20**(1), 171–177.

Barabási, A.-L., Szépfalusy, P., and Vicsek, T. 1991. Multifractal spectra of multi-affine functions. *Physica A*, **178**(1), 17–28.

Bardet, J. M., and Kammoun, I. 2008. Asymptotic properties of the detrended fluctuation analysis of long-range-dependent processes. *IEEE Trans. Inform. Theory*, **54**(5), 2041–2052.

Barral, J., and Mandelbrot, B. B. 2002. Multifractal products of cylindrical pulses. *Probab. Theory Related Fields*, **124**(3), 409–430.

Bashan, A., Bartsch, R., Kantelhardt, J. W., and Havlin, S. 2008. Comparison of detrending methods for fluctuation analysis. *Physica A*, **387**(21), 5080–5090.

Batchelor, G. K. 1953. *The theory of homogeneous turbulence*. Cambridge University Press.

Batchelor, G. K., and Townsend, A. A. 1949. The nature of turbulent motion at large wavenumbers. *Proc. R. Soc. A*, **199**(1057), 238–255.

Battjes, J. A. 1988. Surf zone dynamics. *Annu. Rev. Fluid Mech.*, **20**(1), 257–291.

Bec, J., Biferale, L., Cencini, M., Lanotte, A. S., and Toschi, F. 2006. Effects of vortex filaments on the velocity of tracers and heavy particles in turbulence. *Phys. Fluids*, **18**, 081702.

Bec, J., Homann, H., and Krstulovic, G. 2014. Clustering, fronts, and heat transfer in turbulent suspensions of heavy particles. *Phys. Rev. Lett.*, **112**, 234503.

Beck, C. 2007. Statistics of three-dimensional Lagrangian turbulence. *Phys. Rev. Lett.*, **98**(6), 064502.

Benzi, R., Paladin, G., Vulpiani, A., and Parisi, G. 1984. On the multifractal nature of fully developed turbulence and chaotic systems. *J. Phys. A*, **17**, 3521–3531.

Benzi, R., Ciliberto, S., Tripiccione, R., Baudet, C., Massaioli, F., and Succi, S. 1993a. Extended self-similarity in turbulent flows. *Phys. Rev. E*, **48**(1), 29–32.

Benzi, R., Biferale, L., Crisanti, A., Paladin, G., Vergassola, M., and Vulpiani, A. 1993b. A random process for the construction of multiaffine fields. *Physica D*, **65**(4), 352–358.

Benzi, R., Ciliberto, S., Baudet, C., and Chavarria, G. R. 1995. On the scaling of three-dimensional homogeneous and isotropic turbulence. *Physica D*, **80**(4), 385–398.

Benzi, R., Biferale, L., Calzavarini, E., Lohse, D., and Toschi, F. 2009. Velocity-gradient statistics along particle trajectories in turbulent flows: the refined similarity hypothesis in the Lagrangian frame. *Phys. Rev. E*, **80**(6), 066318.

Beran, J. 1994. *Statistics for long-memory processes*. CRC Press.

Berg, J., Ott, S., Mann, J., and Lüthi, B. 2009. Experimental investigation of Lagrangian structure functions in turbulence. *Phys. Rev. E*, **80**(2), 026316.

Bernard, D. 2000. Influence of friction on the direct cascade of the 2d forced turbulence. *Europhys. Lett.*, **50**, 333–339.

Bernard, P. S., and Wallace, J. M. 2002. *Turbulent flow: analysis, measurement, and prediction*. John Wiley & Sons.

Biagini, F., Hu, Y., Oksendal, B., and Zhang, T. 2008. *Stochastic calculus for fractional Brownian motion and applications*. Springer Verlag.

Biferale, L., Boffetta, G., Celani, A., Crisanti, A., and Vulpiani, A. 1998. Mimicking a turbulent signal: sequential multiaffine processes. *Phys. Rev. E*, **57**, R6261–R6264.

Biferale, L., Cencini, M., Lanotte, A. S., and Vergni, D. 2003. Inverse velocity statistics in two-dimensional turbulence. *Phys. Fluids*, **15**(4), 1012–1020.

Biferale, L., Boffetta, G., Celani, A., Devenish, B. J., Lanotte, A., and Toschi, F. 2004. Multifractal statistics of Lagrangian velocity and acceleration in turbulence. *Phys. Rev. Lett.*, **93**(6), 064502.

References

Blain, S., Guillou, J., Treguer, P., Woerther, P., Delauney, L., Follenfant, E., Gontier, O., Hamon, M., Leilde, B., Masson, A., Tartub, C., and Vuillemin, R. 2004. High frequency monitoring of the coastal environment using the marel buoy. *J. Environ. Monit.*, **6**, 569–575.

Boffetta, G. 2007. Energy and enstrophy fluxes in the double cascade of two-dimensional turbulence. *J. Fluid Mech.*, **589**, 253–260.

Boffetta, G., and Ecke, R. E. 2012. Two-Dimensional Turbulence. *Annu. Rev. Fluid Mech.*, **44**, 427–451.

Boffetta, G., and Musacchio, S. 2010. Evidence for the double cascade scenario in two-dimensional turbulence. *Phys. Rev. E*, **82**(1), 016307.

Boffetta, G., Celani, A., Musacchio, S., and Vergassola, M. 2002. Intermittency in two-dimensional Ekman-Navier-Stokes turbulence. *Phys. Rev. E*, **66**(2), 026304.

Bolgiano, R. 1959. Turbulent spectra in a stably stratified atmosphere. *J. Geophys. Res.*, **64**, 2226–2229.

Boettcher, F., Barth, S., and Peinke, J. 2007. Small and large scale fluctuations in atmospheric wind speeds. *Stoch. Env. Res. Risk A.*, **21**(3), 299–308.

Bouchet, F., and Venaille, A. 2012. Statistical mechanics of two-dimensional and geophysical flows. *Phys. Rep.*, **515**, 227–295.

Brown, E., and Ahlers, G. 2007. Large-scale circulation model for turbulent Rayleigh-Bénard convection. *Phys. Rev. Lett.*, **98**(Mar.), 134501.

Brown, E., Nikolaenko, A., and Ahlers, G. 2005. Reorientation of the large-scale circulation in turbulent Rayleigh-Bénard convection. *Phys. Rev. Lett.*, **95**, 084503.

Calif, R., and Schmitt, F. G. 2012. Modeling of atmospheric wind speed sequence using a lognormal continuous stochastic equation. *J. Wind Eng. Ind. Aerodyn.*, **109**, 1–8.

Calif, R., and Schmitt, F. G. 2014. Multiscaling and joint multiscaling description of the atmospheric wind speed and the aggregate power output from a wind farm. *Nonlinear Proc. Geoph.*, **21**(2), 379–392.

Calif, R., Schmitt, F. G., and Huang, Y. 2013. Multifractal description of wind power fluctuations using arbitrary order Hilbert spectral analysis. *Physica A*, **392**, 4106–4120.

Celani, A., Lanotte, A., Mazzino, A., and Vergassola, M. 2000. Universality and saturation of intermittency in passive scalar turbulence. *Phys. Rev. Lett.*, **84**, 2385–2388.

Celani, A., Musacchio, S., and Vincenzi, D. 2010. Turbulence in more than two and less than three dimensions. *Phys. Rev. Lett.*, **104**(18), 184506.

Chainais, P. 2006. Multidimensional infinitely divisible cascades. *EPJ B*, **51**(2), 229–243.

Champagne, F. H. 1978. The fine-scale structure of the turbulent velocity field. *J. Fluid Mech.*, **86**(01), 67–108.

Chang, G. C., and Dickey, T. D. 2001. Optical and physical variability on timescales from minutes to the seasonal cycle on the New England shelf: July 1996 to June 1997. *J. Geophys. Res.*, **106**, 9435–9453.

Chavez, F. P., Pennington, J., Herlien, R., Jannasch, H., Thurmond, G., and Friederich, G. E. 1997. Moorings and drifters for real-time interdisciplinary oceanography. *J. Atmos. Oceanic Technol.*, **14**, 1199–1211.

Chen, J., Xu, Y. L., and Zhang, R. C. 2004. Modal parameter identification of Tsing Ma suspension bridge under Typhoon Victor: EMD-HT method. *J. Wind Eng. Ind. Aerodyn.*, **92**(10), 805–827.

Chen, S. Y., and Cao, N. 1995. Inertial range scaling in turbulence. *Phys. Rev. E*, **52**(6), 5757–5759.

Chen, S. Y., Sreenivasan, K. R., Nelkin, M., and Cao, N. Z. 1997. Refined similarity hypothesis for transverse structure functions in fluid turbulence. *Phys. Rev. Lett.*, **79**(12), 2253–2256.

Chen, S. Y., Ecke, R. E., Eyink, G. L., Wang, X., and Xiao, Z. 2003. Physical mechanism of the two-dimensional enstrophy cascade. *Phys. Rev. Lett.*, **91**(21), 214501.

Chen, S. Y., Ecke, R. E., Eyink, G. L., Rivera, M., Wan, M., and Xiao, Z. 2006. Physical mechanism of the two-dimensional inverse energy cascade. *Phys. Rev. Lett.*, **96**(8), 84502.

Chen, Z., Ivanov, P. Ch., Hu, K., and Stanley, H. E. 2002. Effect of nonstationarities on detrended fluctuation analysis. *Phys. Rev. E*, **65**(4), 041107.

Chevillard, L., and Meneveau, C. 2006. Lagrangian dynamics and statistical geometric structure of turbulence. *Phys. Rev. Lett.*, **97**(17), 174501.

Chevillard, L., Roux, S. G., Lévêque, E., Mordant, N., Pinton, J.-F., and Arnéodo, A. 2003. Lagrangian velocity statistics in turbulent flows: Effects of dissipation. *Phys. Rev. Lett.*, **91**(21), 214502.

Ching, E. S. C., Chui, K.-W., Shang, X. D., Qiu, X.-L., Tong, P., and Xia, K.-Q. 2004. Velocity and temperature cross-scaling in turbulent thermal convection. *J. Turbul.*, **5**, 027.

Cioni, S., Ciliberto, S., and Sommeria, J. 1995. Temperature structure functions in turbulent convection at low Prandtl number. *Europhys. Lett.*, **32**, 413.

Cohen, L. 1995. *Time-frequency analysis*. Englewood Cliffs, NJ: Prentice Hall PTR.

Collet, P., and Koukiou, F. 1992. Large deviations for multiplicative chaos. *Commun. Math. Phys.*, **147**(2), 329–342.

Corrsin, S. 1951. On the spectrum of isotropic temperature fluctuations in an isotropic turbulence. *J. Appl. Phys.*, **22**, 469.

Corrsin, S. 1975. Limitations of gradient transport models in random walks and in turbulence. *Adv. Geophys.*, **18**, 25–60.

Costa, A., Crespo, A., Navarro, J., Lizcano, G., Madsen, H., and Feitosa, E. 2008. A review on the young history of the wind power short-term prediction. *Renew. Sust. Eenerg. Rev.*, **12**(6), 1725–1744.

Coughlin, K. T., and Tung, K. K. 2004. 11-Year solar cycle in the stratosphere extracted by the empirical mode decomposition method. *Adv. Space Res.*, **34**(2), 323–329.

Dahlstedt, K., and Jensen, H. J. 2005. Fluctuation spectrum and size scaling of river flow and level. *Physica A*, **348**, 596–610.

Daubechies, I. 1992. *Ten lectures on wavelets*. Philadelphia: SIAM.

Davidson, P. A., and Pearson, B. R. 2005. Identifying turbulent energy distribution in real, rather than Fourier, space. *Phys. Rev. Lett.*, **95**, 214501.

Davis, A., Marshak, A., Wiscombe, W., and Cahalan, R. 1994. Multifractal characterizations of nonstationarity and intermittency in geophysical fields: observed, retrieved, or simulated. *Journal of Geophysical Research: Atmospheres*, **99**(D4), 8055–8072.

Davis, A. B., Marshak, A. L., and Wiscombe, W. J. 1993. Bi-multifractal analysis and multi-affine modeling of non-stationary geophysical processes, application to turbulence and clouds. *Fractals*, **1**(3), 560–567.

de Montera, L., Jouini, M., Verrier, S., Thiria, S., and Crepon, M. 2011. Multifractal analysis of oceanic chlorophyll maps remotely sensed from space. *Ocean Sci.*, **7**(2), 219–229.

de Montera, L., Barthès, L., Mallet, C., and Golé, P. 2009. The effect of rain–no rain intermittency on the estimation of the universal multifractals model parameters. *J. Hydrometeor.*, **10**(2), 493–506.

Denman, K., Okubo, A., and Platt, T. 1977. The chlorophyll fluctuation spectrum in the sea. *Limnol. Oceanogr.*, **22**(6), 1033–1038.

Denman, K. L. 1976. Covariability of chlorophyll and temperature in the sea. *Deep Sea Res.: Oceanogr. Abstr.*, **23**, 539–550.

Desprez, M. 2000. Physical and biological impact of marine aggregate extraction along the French coast of the Eastern English Channel: short-and long-term post-dredging restoration. *ICES Journal of Marine Science: Journal du Conseil*, **57**(5), 1428–1438.

Dickey, T. D. 1991. The emergence of concurrent high resolution physical and bio-optical measurements in the upper ocean and their applications. *Rev. Geophys.*, **29**, 383–413.

Dubrulle, B. 1994. Intermittency in fully developed turbulence: Log-Poisson statistics and generalized scale covariance. *Phys. Rev. Lett.*, **73**(7), 959–962.

Dur, G., Schmitt, F. G., and Souissi, S. 2007. Analysis of high frequency temperature time series in the Seine estuary from the Marel autonomous monitoring buoy. *Hydrobiologia*, **588**(1), 59–68.

Echeverria, J. C., Crowe, J. A., Woolfson, M. S., and Hayes-Gill, B. R. 2001. Application of empirical mode decomposition to heart rate variability analysis. *Med. Biol. Eng. Comput.*, **39**(4), 471–479.

Eggers, J., and Grossmann, S. 1992. Effect of dissipation fluctuations on anomalous velocity scaling in turbulence. *Phys. Rev. A*, **45**(4), 2360.

Egolf, P. W., and Weiss, D. A. 1998. Difference-quotient turbulence model: the axisymmetric isothermal jet. *Phys. Rev. E*, **58**(1), 459.

Embrechts, P., and Maejima, M. 2002. *Self-similar processes*. Princeton University Press.

Falkovich, G., and Lebedev, V. 1994. Universal direct cascade in two-dimensional turbulence. *Phys. Rev. E*, **50**(5), 3883.

Falkovich, G., and Lebedev, V. 2011. Vorticity statistics in the direct cascade of two-dimensional turbulence. *Phys. Rev. E*, **83**(4), 045301.

Falkovich, G., and Sreenivasan, K. R. 2006. Lessons from hydrodynamic turbulence. *Phys. Today*, **59**, 43.

Falkovich, G., Gawedzki, K., and Vergassola, M. 2001. Particles and fields in fluid turbulence. *Rev. Mod. Phys.*, **73**(4).

Falkovich, G., Xu, H. T., Pumir, A., Bodenschatz, E., Biferale, L., Boffetta, G., Lanotte, A.S., and Toschi, F. 2012. On Lagrangian single-particle statistics. *Phys. Fluids*, **24**(4), 055102.

Farge, M. 1992. Wavelet transforms and their applications to turbulence. *Annu. Rev. Fluid Mech.*, **24**(1), 395–457.

Farge, M., Kevlahan, N., Perrier, V., and Goirand, E. 1996. Wavelets and turbulence. *Proc. IEEE*, **84**(4), 639–669.

Feller, W. 1971. *An introduction to probalitity theory and its applications*. New York: Wiley.

Flandrin, P. 1998. *Time-frequency/time-scale analysis*. San Diego, CA: Academic Press.

Flandrin, P., and Gonçalvès, P. 2004. Empirical mode decompositions as data-driven wavelet-like expansions. *IJWMIP*, **2**(4), 477–496.

Flandrin, P., Rilling, G., and Gonçalvès, P. 2004. Empirical mode decomposition as a filter bank. *IEEE Signal Processing Lett.*, **11**(2), 112–114.

Frisch, U. 1995. *Turbulence: the legacy of AN Kolmogorov*. Cambridge University Press.

Frisch, U., and Matsumoto, T. 2002. On multifractality and fractional derivatives. *J. Stat. Phys.*, **108**(5-6), 1181–1202.

Frisch, U., Sulem, P. L., and Nelkin, M. 1978. A simple dynamical model of intermittent fully developed turbulence. *J. Fluid Mech.*, **87**(4), 719–736.

Gagne, Y. 1980. *Contribution à l'étude expérimentale de l'intermittence de la turbulence à petite échelle*. PhD thesis.

Garcia, H. E., and Gordon, L. I. 1992. Oxygen solubility in seawater: better fitting equations. *Limnol. Oceanogr.*, **37**, 1307–1312.

Ghashghaie, S., Breymann, W., Peinke, J., Talkner, P., and Dodge, Y. 1996. Turbulent cascades in foreign exchange markets. *Nature*, **381**(6585), 767–770.

Gipe, P. 1995. *Wind energy comes of age*. Vol. 4. New York: John Wiley & Sons.

Gnedenko, B. V., and Kolmogorov, A. N. 1954. *Limit distributions for sums of independent random variables*.

Gottschall, Julia, and Peinke, Joachim. 2008. How to improve the estimation of power curves for wind turbines. *Environ. Res. Lett.*, **3**(1), 015005.

Grant, H. L., Stewart, R. W., and Moilliet, A. 1962. Turbulence spectra from a tidal channel. *J. Fluid Mech.*, **12**(2), 241–268.

Grassberger, P., and Procaccia, I. 1983. Generalized dimensions of strange attractors. *Phys. Rev. Lett.*, **50**(6), 346.

Grossmann, S., and Lohse, D. 2004. Fluctuations in turbulent Rayleigh–Bénard convection: the role of plumes. *Phys. Fluids*, **16**, 4462.

Gurvich, A. S. 1960. Frequency spectra and distribution functions of vertical wind components. *Izvestia ANSSSR Geophys Ser*, **7**, 1042.

Gurvich, A. S., and Yaglom, A. M. 1967. Breakdown of eddies and probability distributions for small-scale turbulence. *Phys. Fluids*, **10**(9), S59–S65.

Gurvich, A. S., and Zubkovskii, S. L. 1963. Experimental estimate of fluctuations in the turbulent energy dissipation. *Izv. Akad. Nauk SSSR, Ser. Geofiz*, **12**, 1856–1858.

Haar, A. 1910. On the theory of orthogonal function systems. *Mathematische Annalen*, **69**, 331–371.

Halsey, T. C., Jensen, M. H., Kadanoff, L. P., Procaccia, I., and Shraiman, B. I. 1986. Fractal measures and their singularities: the characterization of strange sets. *Phys. Rev. A*, **33**(2), 1141–1151.

Hartlep, T., Tilgner, A., and Busse, F. H. 2003. Large scale structures in Rayleigh-Bénard convection at high Rayleigh numbers. *Phys. Rev. Lett.*, **91**(6), 64501.

He, G. W. 2011. Anomalous scaling for Lagrangian velocity structure functions in fully developed turbulence. *Phys. Rev. E*, **83**(2), 025301.

He, G. W., and Zhang, J. B. 2006. Elliptic model for space-time correlations in turbulent shear flows. *Phys. Rev. E*, **73**, 055303(R).

He, X. Z., and Tong, P. 2011. Kraichnan's random sweeping hypothesis in homogeneous turbulent convection. *Phys. Rev. E*, **83**, 037302.

He, X. Z., He, G. W., and Tong, P. 2010. Small-scale turbulent fluctuations beyond Taylor's frozen-flow hypothesis. *Phys. Rev. E*, **81**, 065303(R).

He, X. Z., Funfschilling, D., Nobach, H., Bodenschatz, E., and Ahlers, G. 2012. Transition to the ultimate state of turbulent Rayleigh-Bénard convection. *Phys. Rev. Lett.*, **108**(2), 024502.

Heneghan, C., and McDarby, G. 2000. Establishing the relation between detrended fluctuation analysis and power spectral density analysis for stochastic processes. *Phys. Rev. E*, **62**(5), 6103–6110.

Hentschel, H. G. E., and Procaccia, I. 1983. The infinite number of generalized dimensions of fractals and strange attractors. *Physica D*, **8**(3), 435–444.

Hinze, J. O., Sonnenberg, R. E., and Builtjes, P. J. H. 1974. Memory effect in a turbulent boundary-layer flow due to a relatively strong axial variation of the mean-velocity gradient. *Appl. Sci. Res.*, **29**(1), 1–13.

Hu, K., Ivanov, P. C., Chen, Z., Carpena, P., and Stanley, H. E. 2001. Effect of trends on detrended fluctuation analysis. *Phys. Rev. E*, **64**(1), 11114.

Huang, N. E. 2005. *Hilbert-Huang transform and its applications*. World Scientific. Chap. 1. Introduction to the Hilbert-Huang transform and its related mathematical problems, 1–26.

Huang, N. E., and Wu, Z. 2005. An adaptive data analysis method for nonlinear and nonstationary time series: the empirical mode decomposition and Hilbert spectrum analysis. *Proceedings of the 4th International Conference on Wavelet and Its Application, Macao*.

Huang, N. E., Wu, M. L., Long, S. R., Shen, S. S. P., Qu, W., Gloersen, P., and Fan, K. L. 2003a. A confidence limit for the empirical mode decomposition and Hilbert spectral analysis. *Proc. R. Soc. A*, **459**(2037), 2317–2345.

Huang, N. E., Wu, M. L., Qu, W., Long, S. R., and Shen, S. S. P. 2003b. Applications of Hilbert-Huang transform to non-stationary financial time series analysis. *Appl. Stoch. Model Bus.*, **19**(3), 245–268.

Huang, N. E., Shen, Z., Long, S. R., Wu, M. C., Shih, H. H., Zheng, Q., Yen, N. C., Tung, C. C., and Liu, H. H. 1998. The empirical mode decomposition and the Hilbert spectrum for nonlinear and non-stationary time series analysis. *Proc. R. Soc. A*, **454**(1971), 903–995.

Huang, N. E., Shen, Z., and Long, S. R. 1999. A new view of nonlinear water waves: the Hilbert spectrum. *Annu. Rev. Fluid Mech.*, **31**(1), 417–457.

Huang, Y. X. 2009. *Arbitrary-order Hilbert spectral analysis: definition and application to fully developed turbulence and environmental time series*. PhD thesis, Université des Sciences et Technologies de Lille - Lille 1, France & Shanghai University, China.

Huang, Y. X. 2014. Detrended structure-function in fully developed turbulence. *J. Turbul.*, **15**(4), 209–220.

Huang, Y. X., Schmitt, F. G., Lu, Z. M, and Liu, Y. L. 2008a. An amplitude-frequency study of turbulent scaling intermittency using Hilbert spectral analysis. *Europhys. Lett.*, **84**, 40010.

Huang, Y. X., Schmitt, F. G., Lu, Z. M., and Liu, Y. L. 2008b. Analyse de l'invariance d'échelle de séries temporelles par la décomposition modale empirique et l'analyse spectrale de Hilbert. *Traitement du Signal*, **25**, 481–492.

Huang, Y. X., Schmitt, F. G., Lu, Z. M., and Liu, Y. L. 2009a. Analysis of Daily River Flow Fluctuations Using Empirical Mode Decomposition and Arbitrary Order Hilbert Spectral Analysis. *J. Hydrol.*, **373**, 103–111.

Huang, Y. X., Schmitt, F. G., Lu, Z. M., and Liu, Y. L. 2009b. Autocorrelation function of velocity increments in fully developed turbulence. *EPL*, **86**, 40010.

Huang, Y. X., Schmitt, F. G., Lu, Z. M., Fougairolles, P., Gagne, Y., and Liu, Y. L. 2010. Second-order structure function in fully developed turbulence. *Phys. Rev. E*, **82**(2), 026319.

Huang, Y. X., Schmitt, F. G., Hermand, J.-P., Gagne, Y., Lu, Z. M., and Liu, Y. L. 2011a. Arbitrary-order Hilbert spectral analysis for time series possessing scaling statistics: comparison study with detrended fluctuation analysis and wavelet leaders. *Phys. Rev. E*, **84**(1), 016208.

Huang, Y. X., Schmitt, F. G., Zhou, Q., Qiu, X., Shang, X. D., Lu, Z. M., and Liu, Y. L. 2011b. Scaling of maximum probability density functions of velocity and temperature increments in turbulent systems. *Phys. Fluids*, **23**, 125101.

Huang, Y. X., Biferale, L., Calzavarini, E., Sun, C., and Toschi, F. 2013. Lagrangian single particle turbulent statistics through the Hilbert-Huang transforms. *Phys. Rev. E*, **87**, 041003(R).

Huang, Y. X., Schmitt, F. G., and Gagne, Y. 2014. Two-scale correlation and energy cascade in three-dimensional turbulent flows. *J. Stat. Mech*, **5**, P05002.

Hwang, P. A., Huang, N. E., and Wang, D. W. 2003. A note on analyzing nonlinear and nonstationary ocean wave data. *Appl. Ocean Res.*, **25**(4), 187–193.

Inoue, E. 1952. Turbulent fluctuations in temperature in the atmosphere and oceans. *J. Meteor. Soc. Japan*, **29**, 246–253.

Irion, R. 1999. Soap films reveal whirling worlds of turbulence. *Science*, **284**(5420), 1609–1610.

Jaffard, S. 1999. The multifractal nature of Lévy processes. *Probab. Theory Related Fields*, **114**(2), 207–227.

Jaffard, S., Lashermes, B., and Abry, P. 2007. Wavelet leaders in multifractal analysis. In *Wavelet analysis and applications*. Birkhauser Verlag, Basel: Springer, 201–246.

Janicki, A., and Weron, A. 1994. *Simulation and chaotic behavior of alpha-stable stochastic processes*. New York: Marcel Dekker.

Jánosi, I. M., and Müller, R. 2005. Empirical mode decomposition and correlation properties of long daily ozone records. *Phys. Rev. E*, **71**(5), 56126.

Juneja, A., Lathrop, D. P., Sreenivasan, K. R., and Stolovitzky, G. 1994. Synthetic turbulence. *Phys. Rev. E*, **49**(6), 5179.

Kader, B. A., and Yaglom, A. M. 1984. Turbulent structure of an unstable atmospheric surface layer. In *Nonlinear and Turbulent Processes in Physics*, vol. 1, 829.

Kahane, J. P. 1985. Sur le chaos multiplicatif. *Ann. Sci. Math. Québec*, **9**(2), 105–150.

Kang, H., Chester, S., and Meneveau, C. 2003. Decaying turbulence in an active-grid-generated flow and comparisons with large-eddy simulation. *J. Fluid Mech.*, **480**, 129–160.

Kantelhardt, J. W., Zschiegner, S. A., Koscielny-Bunde, E., Havlin, S., Bunde, A., and Stanley, H. E. 2002. Multifractal detrended fluctuation analysis of nonstationary time series. *Physica A*, **316**(1-4), 87–114.

Katul, G., and Chu, C.-R. 1998. A theoretical and experimental investigation of energy-containing scales in the dynamic sublayer of boundary-layer flows. *Boundary-Layer Meteorol.*, **86**(2), 279–312.

Katul, G. G., Chu, C. R., Parlange, M. B., Albertson, J. D., and Ortenburger, T. A. 1995. Low-wavenumber spectral characteristics of velocity and temperature in the atmospheric surface layer. *J. Geophys. Res.*, **100**(D7), 14243–14255.

Katul, G. G., Porporato, A., and Nikora, V. 2012. Existence of k- 1 power-law scaling in the equilibrium regions of wall-bounded turbulence explained by Heisenberg's eddy viscosity. *Phys. Rev. E*, **86**(6), 066311.

Katzenstein, W., Fertig, E., and Apt, J. 2010. The variability of interconnected wind plants. *Energy Policy*, **38**(8), 4400–4410.

Kellay, H., Wu, X. L., and Goldburg, W. I. 1998. Vorticity measurements in turbulent soap films. *Phys. Rev. Lett.*, **80**(2), 277–280.

Kellay, H., and Goldburg, W. I. 2002. Two-dimensional turbulence: a review of some recent experiments. *Rep. Prog. Phys.*, **65**(5), 845.

Kelley, D. H., and Ouellette, N. T. 2011. Spatiotemporal persistence of spectral fluxes in two-dimensional weak turbulence. *Phys. Fluids*, **23**(11), 115101.

Khurana, N., and Ouellette, N. T. 2012. Interactions between active particles and dynamical structures in chaotic flow. *Phys. Fluids*, **24**(9), 091902.

Kida, S. 1991. Log stable distribution and intermittency of turbulence. *J. Phys. Soc. Jpn.*, **60**(1), 5–8.

Kolmogorov, A. N. 1940. The Wiener spiral and some other interesting curves in Hilbert space. *Dokl. Akad. Nauk SSSR*, **26**(2), 115–118.

Kolmogorov, A. N. 1941a. Energy dissipation in locally isotropic turbulence. *Doklady AN SSSR*, **32**(1), 19–21.

Kolmogorov, A. N. 1941b. Local structure of turbulence in an incompressible fluid at very high Reynolds numbers. *Dokl. Akad. Nauk SSSR*, **30**, 301.

Kolmogorov, A. N. 1962. A refinement of previous hypotheses concerning the local structure of turbulence in a viscous incompressible fluid at high Reynolds number. *J. Fluid Mech.*, **13**, 82–85.

Korotenko, K. A., Sentchev, A. V., and Schmitt, F. G. 2012. Effect of variable winds on current structure and Reynolds stresses in a tidal flow: analysis of experimental data in the eastern English Channel. *Ocean Science*, **8**(6), 1025–1040.

Koscielny-Bunde, E., Kantelhardt, J. W., Braun, P., Bunde, A., and Havlin, S. 2006. Long-term persistence and multifractality of river runoff records: detrended fluctuation studies. *J. Hydrol.*, **322**(1-4), 120–137.

Kraichnan, R. H. 1967. Inertial Ranges in Two-Dimensional Turbulence. *Phys. Fluids*, **10**, 1417–1423.

Kraichnan, R. H., and Montgomery, D. 1980. Two-dimensional turbulence. *Rep. Prog. Phys.*, **43**, 547.

Kunnen, R. P. J., Clercx, H. J. H., Geurts, B. J., van Bokhoven, L. J. A., Akkermans, R. A. D., and Verzicco, R. 2008. Numerical and experimental investigation of structure-function scaling in turbulent Rayleigh-Bénard convection. *Phys. Rev. E*, **77**(1), 016302.

Kuramoto, Y., Battogtokh, D., and Nakao, H. 1998. Multiaffine chemical turbulence. *Phys. Rev. Lett.*, **81**(16), 3543.

Landau, L. D., and Lifshits, E. M. 1944. *Fluid mechanics, 1st Russian edn.*

Landberg, L. 1999. Short-term prediction of the power production from wind farms. *J. Wind Eng. Ind. Aerodyn.*, **80**(1), 207–220.

Lashermes, B., Abry, P., and Chainais, P. 2004. New insights into the estimation of scaling exponents. *IWMIP*, **2**(04), 497–523.

Lashermes, B., Jaffard, S., and Abry, P. 2005. Wavelet leader based multifractal analysis. In *IEEE International Conference on Acoustics, Speech, and Signal Processing, 2005. Proceedings ICASSP'05.*, vol. 4. IEEE, iv–161.

Lashermes, B., Roux, S. G., Abry, P., and Jaffard, S. 2008. Comprehensive multifractal analysis of turbulent velocity using the wavelet leaders. *Eur. Phys. J. B*, **61**(2), 201–215.

Lévy, P. 1937. *Théorie de laddition des variables aléatoires*. Gauthiers-Villars, Paris.

Li, M. Y., and Huang, Y. X. 2014. Hilbert–Huang Transform based multifractal analysis of China stock market. *Physica A*, **406**, 222–229.

Loh, C. H., Wu, T. C., and Huang, N. E. 2001. Application of the Empirical Mode Decomposition-Hilbert Spectrum Method to Identify Near-Fault Ground-Motion Characteristics and Structural Responses. *BSSA*, **91**(5), 1339–1357.

Lohse, D., and Xia, K.-Q. 2010. Small-scale properties of turbulent Rayleigh-Bénard convection. *Annu. Rev. Fluid Mech.*, **42**, 335–364.

Long, S. R., Huang, N. E., Tung, C. C., Wu, M. L., Lin, R. Q., Mollo-Christensen, E., and Yuan, Y. 1995. The Hilbert techniques: an alternate approach for non-steady time series analysis. *IEEE Geoscience and Remote Sensing Soc. Lett.*, **3**, 6–11.

Loutridis, S. J. 2005. Resonance identification in loudspeaker driver units: A comparison of techniques. *Appl. Acoust.*, **66**(12), 1399–1426.

Lovejoy, S, and Schertzer, D. 2012. Haar wavelets, fluctuations and structure functions: convenient choices for geophysics. *Nonlinear Proc. Geoph.*, **19**(5), 513–527.

Lovejoy, S, Schertzer, D, Tessier, Y, and Gaonac'h, H. 2001a. Multifractals and resolution-independent remote sensing algorithms: the example of ocean colour. *Int. J. Remote Sens.*, **22**(7), 1191–1234.

Lovejoy, S., Currie, W. J. S., Tessier, Y., Claereboudt, M. R., Bourget, E., Roff, J. C., and Schertzer, D. 2001b. Universal multifractals and ocean patchiness: phytoplankton, physical fields and coastal heterogeneity. *J. Plankton Res.*, **23**(2), 117–141.

Lovejoy, Shaun, and Schertzer, Daniel. 2013. *The weather and climate: emergent laws and multifractal cascades*. Cambridge University Press.

Ludena, Carenne. 2008. Lp-variations for multifractal fractional random walks. *Ann. Appl. Probab.*, **18**(3), 1138–1163.

Lumley, J. L. 1970. Toward a turbulent constitutive relation. *J. Fluid Mech.*, **41**(02), 413–434.

Maejima, M. 1983. On a class of self-similar processes. *Probab. Theory Related Fields*, **62**(2), 235–245.

Malik, S. C., and Arora, S. 1992. *Mathematical Analysis*. New York: John Wiley & Sons Inc.

Mallat, S., and Hwang, W. L. 1992. Singularity detection and processing with wavelets. *IEEE Trans. Inform. Theory*, **38**(2), 617–643.

Mallat, S. G. 1999. *A wavelet tour of signal processing*. Burlington: Academic Press.

Mandelbrot, B. B. 1974. Intermittent turbulence in self-similar cascades: divergence of high moments and dimension of the carrier. *J. Fluid Mech.*, **62**(2), 331–358.

Mandelbrot, B. B. 1983. *The fractal geometry of nature/Revised and enlarged edition*. Vol. 1.

Mandelbrot, B. B. 1991. Random multifractals: negative dimensions and the resulting limitations of the thermodynamic formalism. *Proc. R. Soc. A*, **434**(1890), 79–88.

Mandelbrot, B. B., Fisher, A. J., and Calvet, L. E. 1997. A Multifractal Model of Assets Returns. Cowles Foundation discussion paper no. 1164.

Mandelbrot, B. B., and Van Ness, J.W. 1968. Fractional Brownian Motions, Fractional Noises and Applications. *SIAM Review*, **10**, 422.

Manneville, Paul. 2004. *Instabilités, chaos et turbulence*. Editions Ecole Polytechnique.

Mantegna, R. N., and Stanley, H. E. 1996. Turbulence and financial markets. *Nature*, **383**(6601), 587–588.

Marsan, D., Schertzer, D., and Lovejoy, S. 1996. Causal space-time multifractal processes: Predictability and forecasting of rain fields. *Journal of Geophysical Research: Atmospheres*, **101**(D21), 26333–26346.

Meneveau, C. 2011. Lagrangian dynamics and models of the velocity gradient tensor in turbulent flows. *Annu. Rev. Fluid Mech.*, **43**, 219–245.

Meneveau, C., and Sreenivasan, K. R. 1987. Simple multifractal cascade model for fully developed turbulence. *Phys. Rev. Lett.*, **59**(13), 1424.

Merrifield, S. T., Kelley, D. H., and Ouellette, N. T. 2010. Scale-dependent statistical geometry in two-dimensional flow. *Phys. Rev. Lett.*, **104**(25), 254501.

Meyer, Yves. 1995. *Wavelets and operators*. Vol. 1. Cambridge: Cambridge University Press.

Milan, P., Wächter, M., and Peinke, J. 2013. Turbulent character of wind energy. *Phys. Rev. Lett.*, **110**(13), 138701.

Molla, K. I., Rahman, M. S., Sumi, A., and Banik, P. 2006. Empirical mode decomposition analysis of climate changes with special reference to rainfall data. *Discrete Dyn. Nat. Soc.*, 45348.

Monin, A. S., and Yaglom, A. M. 1971. *Statistical Fluid Mechanics vd II*. MIT Press.

Morales, A, Wächter, M, and Peinke, J. 2012. Characterization of wind turbulence by higher-order statistics. *Wind Energy*, **15**(3), 391–406.

Mordant, N., Delour, J., Léveque, E., Arnéodo, A., and Pinton, J.-F. 2002. Long time correlations in Lagrangian dynamics: a key to intermittency in turbulence. *Phys. Rev. Lett.*, **89**(25), 254502.

Muzy, J. F., Bacry, E., and Arneodo, A. 1991. Wavelets and multifractal formalism for singular signals: application to turbulence data. *Phys. Rev. Lett.*, **67**(25), 3515–3518.

Muzy, J. F., Bacry, E., and Arneodo, A. 1993. Multifractal formalism for fractal signals: the structure-function approach versus the wavelet-transform modulus-maxima method. *Phys. Rev. E*, **47**(2), 875–884.

Muzy, J.-F., Bacry, E., and Kozhemyak, A. 2006. Extreme values and fat tails of multifractal fluctuations. *Phys. Rev. E*, **73**(6), 066114.

Muzy, J.-F., and Bacry, E. 2002. Multifractal stationary random measures and multifractal random walks with log infinitely divisible scaling laws. *Phys. Rev. E*, **66**(5), 056121.

Nakao, H. 2000. Multi-scaling properties of truncated Lévy flights. *Phys. Lett. A*, **266**(4), 282–289.

Nam, K., Ott, E., Antonsen Jr, T. M., and Guzdar, P. N. 2000. Lagrangian chaos and the effect of drag on the enstrophy cascade in two-dimensional turbulence. *Phys. Rev. Lett.*, **84**(22), 5134–5137.

Nam, S., Kim, G., Kim, K. R., Kim, K., Cheng, L. Oh, Kim, K. W., Ossi, H., and Kim, Y. G. 2005. Application of real-time monitoring buoy systems for physical and biogeochemical parameters in the coastal ocean around the Korean peninsula. *Mar. Technol. Soc. J.*, **39**(2), 70–80.

Nickels, T. B. B., Marusic, I., Hafez, S., and Chong, M. S. 2005. Evidence of the k_1^{-1} Law in a High-Reynolds-Number Turbulent Boundary Layer. *Phys. Rev. Lett.*, **95**(7), 074501.

Nicolis, G., and Nicolis, C. 2012. *Foundations of complex systems: emergence, information and prediction*. World Scientific.

Niemela, J. J., Skrbek, L., Sreenivasan, K. R., and Donnelly, R. J. 2000. Turbulent convection at very high Rayleigh numbers. *Nature*, **404**(6780), 837–840.

Nieves, V., Llebot, C., Turiel, A., Solé, J.ordi, García-Ladona, E., Estrada, M., and Blasco, D. 2007. Common turbulent signature in sea surface temperature and chlorophyll maps. *Geophys. Res. Lett.*, **34**(23).

Nikias, C. L, and Shao, M. 1995. *Signal processing with alpha-stable distributions and applications*. Wiley-Interscience.

Nikora, V. 1999. Origin of the -1 spectral law in wall-bounded turbulence. *Phys. Rev. Lett.*, **83**(4), 734.

Novikov, E. A. 1969. Scale similarity for random fields. *Soviet Physics Doklady*, **14**, 104–107.

Novikov, E. A. 1971. Intermittency and scale similarity in the structure of a turbulent flow. *J. Appl. Math. Mech.*, **35**(2), 231–241.

Novikov, E. A. 1989. Two-particle description of turbulence, Markov property, and intermittency. *Phys. Fluids A*, **1**(2), 326–330.

Novikov, E. A. 1990. The effects of intermittency on statistical characteristics of turbulence and scale similarity of breakdown coefficients. *Phys. Fluids A*, **2**(5), 814–820.

Novikov, E. A. 1994. Infinitely divisible distributions in turbulence. *Phys. Rev. E*, **50**(5), R3303.

Novikov, E. A., and Stewart, R. W. 1964. The intermittency of turbulence and the spectrum of energy dissipation fluctuations. *Bull. Acad. Sci. SSSR Geophy. Ser.*, **3**, 408–413.

Obukhov, A. M. 1941. Spectral energy distribution in a turbulent flow. *Dokl. Akad. Nauk SSSR*, **32**, 22–24.

Obukhov, A. M. 1949. Structure of the temperature field in a turbulent flow. *Izv. Akad. Nauk SSSR Ser. Geog. i Geofiz.*, **13**, 58–69.

Obukhov, A. M. 1959. On the influence of Archimedean forces on the structure of the temperature field in a turbulent flow. *Doklady Akademi Nauk SSSR*, **125**, 1246–48.

Obukhov, A. M. 1962. Some specific features of atmospheric turbulence. *J. Fluid Mech.*, **13**(1), 77–81.

Oświęcimka, P., Kwapień, J., and Drożdż, S. 2006. Wavelet versus detrended fluctuation analysis of multifractal structures. *Phys. Rev. E*, **74**(1), 16103.

Panchev, S. 1972. *Random Functions and Turbulence*. Oxford: Pergamon.

Paret, J., Jullien, M.C., and Tabeling, P. 1999. Vorticity statistics in the two-dimensional enstrophy cascade. *Phys. Rev. Lett.*, **83**(17), 3418–3421.

Parisi, G., and Frisch, U. 1985. On the singularity spectrum of fully developed turbulence. in M. Ghil, R. Benzi, G. Parisi (eds.), *Turbulence and Predictability in Geophysical Fluid Dynamics and Climatic Dynamics*, Amsterdam: North-Holland, 84–87.

Pecknold, S., Lovejoy, S., Schertzer, D., Hooge, C., and Malouin, J. F. 1993. The simulation of universal multifractals. In Perdang, J. M., and A., Lejeune (eds), *Cellular Automata: Prospects in astrophysical applications*, vol. 1. World Scientific, 228–267.

Peinke, J., Barth, S., Boettcher, F., Heinemann, D., and Lange, B. 2004. Turbulence, a challenging problem for wind energy. *Physica A*, **338**(1), 187–193.

Peng, C. K., Buldyrev, S. V., Havlin, S., Simons, M., Stanley, H. E., and Goldberger, A. L. 1994. Mosaic organization of DNA nucleotides. *Phys. Rev. E*, **49**(2), 1685–1689.

Percival, D. B., and Walden, A. T. 1993. *Spectral Analysis for Physical Applications: Multitaper and Conventional Univariate Techniques*. Cambridge: Cambridge University Press.

Perpete, N. 2013. Construction of multifractal fractional randowm walks with Hurst index smaller than 1/2. *Stoch. Dyn.*, **13**(4), 1350003.

Perpete, N., and Schmitt, F. G. 2011. A discrete log-normal process to sequentially generate a multifractal time series. *J. Stat. Mech.*, **2011**(12), P12013.

Perry, A. E., Henbest, S., and Chong, M. S. 1986. A theoretical and experimental study of wall turbulence. *J. Fluid Mech.*, **165**, 163–199.

Pipiras, V., and Taqqu, M. S. 2000. Integration questions related to fractional Brownian motion. *Probab. Theory Related Fields*, **118**(2), 251–291.

Piquet, J. 1999. *Turbulent flows: models and physics*. Berlin: Springer.

Platt, T., and Denman, K. L. 1975. Spectral analysis in ecology. *Annu. Rev. Ecol. Syst.*, 189–210.

Pond, S., and Stewart, R. W. 1965. Measurements of the statistical characteristics of small-scale turbulent motions. *Izv. Atmos. Oceanic Phys*, **1**, 914–919.

Ponomarenko, V. I., Prokhorov, M. D., Bespyatov, A. B., Bodrov, M. B., and Gridnev, V. I. 2005. Deriving main rhythms of the human cardiovascular system from the heartbeat time series and detecting their synchronization. *Chaos Soliton Fract.*, **23**, 1429–1438.

Pope, S. B. 2000. *Turbulent Flows*. Cambridge: Cambridge University Press.

Pottier, C., Turiel, A., and Garçon, V. 2008. Inferring missing data in satellite chlorophyll maps using turbulent cascading. *Remote Sens. Environ.*, **112**(12), 4242–4260.

Pugh, D. 2004. *Changing sea levels: effects of tides, weather and climate*. Cambridge: Cambridge University Press.

Qiu, X., Mompean, G., Schmitt, F. G., and Thompson, R. L. 2011. Modeling turbulent-bounded flow using non-Newtonian viscometric functions. *J. Turbul* **12**(15), 1–18.

Qiu, X., Huang, Y. X., Zhou, Q., and Sun, C. 2014. Scaling of maximum probability density function of velocity increments in turbulent Rayleigh-Bénard convection. *J. Hydrodyn.*, **26**(3), 351–362.

Rajput, B. S., and Rosinski, J. 1989. Spectral representations of infinitely divisible processes. *Probab. Theory Related Fields*, **82**(3), 451–487.

Renosh, P. R., Schmitt, F. G., Loisel, H., Sentchev, A., and Mériaux, X. 2014. High frequency variability of particle size distribution and its dependency on turbulence over the sea bottom during re-suspension processes. *Cont. Shelf Res.*, **77**, 51–60.

Renosh, P. R., Schmitt, F. G., and Loisel, H. 2015. Scaling analysis of ocean surface turbulent heterogeneities from satellite remote sensing: use of 2D structure functions. *PLoS One*, **10**(5), e0126975.

Rhodes, R., and Vargas, V. 2014. Gaussian multiplicative chaos and applications: a review. *Probability Surveys*, **11**, 315–392.

Richardson, L. F. 1922. *Weather prediction by numerical process*. Cambridge: Cambridge University Press.

Rilling, G., Flandrin, P., and Gonçalvès, P. 2003. On empirical mode decomposition and its algorithms. *IEEE-EURASIP Workshop on Nonlinear Signal and Image Processing*, **3**, 8–11.

Robert, R., and Vargas, V. 2010. Gaussian multiplicative chaos revisited. *Ann. Probab.*, **38**(2), 605–631.

Rodrigues Neto, C., Zanandrea, A., Ramos, F. M., Rosa, R. R., Bolzan, M. J. A., and Sá, L. D. A. 2001. Multiscale analysis from turbulent time series with wavelet transform. *Physica A*, **295**(1-2), 215–218.

Sadegh Movahed, M., Jafari, G. R., Ghasemi, F., Rahvar, S., and Rahimi Tabar, M. R. 2006. Multifractal detrended fluctuation analysis of sunspot time series. *J. Stat. Mech.*, 02003.

Saito, Y. 1992. Log-gamma distribution model of intermittency in turbulence. *J. Phys. Soc. Jpn.*, **61**(2), 403–406.

Samorodnitsky, G., and Taqqu, M. S. 1994. *Stable non-Gaussian random processes: stochastic models with infinite variance*. Chapman & Hall.

Sanchez, I. 2006. Short-term prediction of wind energy production. *Int. J. Forecasting*, **22**(1), 43–56.

Sawford, B. L., and Yeung, P. K. 2011. Kolmogorov similarity scaling for one-particle Lagrangian statistics. *Phys. Fluids*, **23**, 091704.

Schertzer, D., and Lovejoy, S. 1984. On the dimension of atmospheric motions. 505–512.

Schertzer, D., and Lovejoy, S. 1987. Physical modeling and analysis of rain and clouds by anisotropic scaling multiplicative processes. *J. Geophys. Res*, **92**(D8), 9693–9714.

Schertzer, D., and Lovejoy, S. 1992. Hard and soft multifractal processes. *Physica A*, **185**(1), 187–194.

Schertzer, D., Lovejoy, S., and Schmitt, F. G. 1995. Structures in turbulence and multifractal universality. in M. Meneguzzi, A. Pouquet and P. L. Sulem (eds.), *Small-scale structures in 3D hydro and MHD turbulence*. Berlin: Springer Verlag, 137–144.

Schertzer, D., Lovejoy, S., Schmitt, F. G., Chigirinskaya, Y., and Marsan, D. 1997. Multifractal cascade dynamics and turbulent intermittency. *Fractals*, **5**(3), 427–471.

Schmitt, F., Schertzer, D., Lovejoy, S., Brunet, Y., et al. 1994. Empirical study of multifractal phase transitions in atmospheric turbulence. *Nonlinear Proc. Geoph.*, **1**(2/3), 95–104.

Schmitt, F., Schertzer, D., Lovejoy, S., and Brunet, Y. 1996. Multifractal temperature and flux of temperature variance in fully developed turbulence. *Europhys. Lett.*, **34**(3), 195.

Schmitt, F. G. 2007a. About Boussinesq's turbulent viscosity hypothesis: historical remarks and a direct evaluation of its validity. *C.R. Mécanique*, **335**(9), 617–627.

Schmitt, F. G. 2007b. Direct test of a nonlinear constitutive equation for simple turbulent shear flows using DNS data. *Commun. Nonlinear Sci. Numer. Simul.*, **12**(7), 1251–1264.

Schmitt, F. G., and Chainais, P. 2007. On causal stochastic equations for log-stable multiplicative cascades. *Eur. Phys. J. B*, **58**(2), 149–158.

Schmitt, F. G., and Seuront, L. 2001. Multifractal random walk in copepod behavior. *Physica A*, **301**(1-4), 375–396.

Schmitt, F. G., Lavallee, D., Schertzer, D., and Lovejoy, S. 1992. Empirical determination of universal multifractal exponents in turbulent velocity fields. *Phys. Rev. Lett.*, **68**(3), 305–308.

Schmitt, F. G., Schertzer, D., and Lovejoy, S. 1999. Multifractal analysis of foreign exchange data. *Appl. Stoch. Mod. Data Anal.*, **15**(1), 29–53.

Schmitt, F. G., Huang, Y. X., Lu, Z., Zongo, S. B., Molinero, J. C., and Liu, Y. 2007. Analysis of nonliner biophysical time series in aquatic environments: scaling properties and empirical mode decomposition. In Tsonis, A., and Elsner, J. (eds.), *Nonlinear Dynamics in Geosciences*. Springer, 261–280.

Schmitt, F. G., Dur, G., Souissi, S., and Brizard Zongo, S. 2008. Statistical properties of turbidity, oxygen and pH fluctuations in the Seine river estuary (France). *Physica A*, **387**(26), 6613–6623.

Schmitt, F. G., Vinkovic, I., and Buffat, M. 2010. Use of Lagrangian statistics for the analysis of the scale separation hypothesis in turbulent channel flow. *Phys. Lett. A*, **374**(33), 3319–3327.

Schmitt, F. G. 2005. Relating Lagrangian passive scalar scaling exponents to Eulerian scaling exponents in turbulence. *EPJ B*, **48**(1), 129–137.

Schmitt, F. G. 2006. Linking Eulerian and Lagrangian structure functions scaling exponents in turbulence. *Physica A*, **368**(2), 377–386.

Schmitt, F. G., and Marsan, D. 2001. Stochastic equations generating continuous multiplicative cascades. *EPJ B*, **20**(1), 3–6.

Schmitt, F. G., Huang, Y. X., Lu, Z. M., Liu, Y. L., and Fernandez, N. 2009. Analysis of velocity fluctuations and their intermittency properties in the surf zone using empirical mode decomposition. *J. Mar. Syst.*, **77**, 473–481.

Serrano, E., and Figliola, A. 2009. Wavelet leaders: a new method to estimate the multifractal singularity spectra. *Physica A*, **388**(14), 2793–2805.

Seuront, L., and Schmitt, F. G. 2005. Multiscaling statistical procedures for the exploration of biophysical couplings in intermittent turbulence. Part I. Theory. *Deep Sea Res. Part II*, **52**(9-10), 1308–1324.

Seuront, L., Schmitt, F., Lagadeuc, Y., Schertzer, D., Lovejoy, S., and Frontier, S. 1996a. Multifractal analysis of phytoplankton biomass and temperature in the ocean. *Geophys. Res. Lett.*, **23**(24), 3591–3594.

Seuront, L., Schmitt, F., Schertzer, D., Lagadeuc, Y., and Lovejoy, S. 1996b. Multifractal intermittency of Eulerian and Lagrangian turbulence of ocean temperature and plankton fields. *Nonlinear Proc. Geoph.*, **3**(4), 236–246.

Seuront, L., Schmitt, F., Lagadeuc, Y., Schertzer, D., and Lovejoy, S. 1999. Universal multifractal analysis as a tool to characterize multiscale intermittent patterns: example of phytoplankton distribution in turbulent coastal waters. *J. Plankton Res.*, **21**(5), 877–922.

Shang, X. D., Qiu, X.-L., Tong, P., and Xia, K.-Q. 2003. Measured local heat transport in turbulent Rayleigh-Bénard convection. *Phys. Rev. Lett.*, **90**, 074501.

Shang, X. D., Qiu, X.-L., Tong, P., and Xia, K.-Q. 2004. Measurements of the local convective heat flux in turbulent Rayleigh-Bénard convection. *Phys. Rev. E*, **70**, 026308.

Shang, X. D., Tong, P., and Xia, K.-Q. 2008. Scaling of the local convective heat flux in turbulent Rayleigh-Bénard convection. *Phys. Rev. Lett.*, **100**(6), 244503.

She, Z. S., and Lévêque, E. 1994. Universal scaling laws in fully developed turbulence. *Phys. Rev. Lett.*, **72**(3), 336–339.

She, Z. S., and Waymire, E. C. 1995. Quantized Energy Cascade and Log-Poisson Statistics in Fully Developed Turbulence. *Phys. Rev. Lett.*, **74**(2), 262–265.

Shraiman, B. I., and Siggia, E. D. 2000. Scalar turbulence. *Nature*, **405**(6787), 639–646.

Siggia, E. D. 1994. High rayleigh number convection. *Annu. Rev. Fluid Mech.*, **26**, 137–168.

Solé, J., Turiel, A., and Llebot, J. E. 2007. Using empirical mode decomposition to correlate paleoclimatic time-series. *Nat. Hazard Earth Sys.*, **7**, 299–307.

Sørensen, P., Hansen, A. D., and Carvalho Rosas, P. A. 2002. Wind models for simulation of power fluctuations from wind farms. *J. Wind Eng. Ind. Aerodyn.*, **90**(12), 1381–1402.

Sreenivasan, K. R. 1991. On Local Isotropy of Passive Scalars in Turbulent Shear Flows. *Proc. R. Soc. A*, **434**(1890), 165–182.

Sreenivasan, K. R., and Antonia, R. A. 1997. The phenomenology of small-scale turbulence. *Annu. Rev. Fluid Mech.*, **29**, 435–472.

Sreenivasan, K. R., and Kailasnath, P. 1993. An update on the intermittency exponent in turbulence. *Phys. Fluids A*, **5**(2), 512–514.

Svendsen, I. A. 1987. Analysis of surf zone turbulence. *J. Geophys. Res.*, **92**(C5), 5115–5124.

Tabeling, P. 2002. Two-dimensional turbulence: a physicist approach. *Phys. Rep.*, **362**(1), 1–62.

Tan, H. S., Huang, Y. X., and Meng, J.-P. 2014. Hilbert Statistics of Vorticity Scaling in Two-Dimensional Turbulence. *Phys. Fluids*, **26**(2), 015106.

Taqqu, M. S., and Wolpert, R. L. 1983. Infinite variance self-similar processes subordinate to a Poisson measure. *Probab. Theory Related Fields*, **62**(1), 53–72.

Taqqu, M. S. 1988. Self-similar processes. *Encyclopedia of Statistical Sciences*.

Taylor, G. I. 1938. The Spectrum of Turbulence. *Proc. R. Soc. A*, **164**(919), 476–490.

Tennekes, H., and Lumley, J. L. 1972. *A First Course in Turbulence*. MIT Press.

Tessier, Y, Lovejoy, S., Schertzer, D., Lavallée, D., and Kerman, B. 1993. Universal multifractal indices for the ocean surface at far red wavelengths. *Geophys. Res. Lett.*, **20**(12), 1167–1170.

Tessier, Y., Lovejoy, S., Hubert, P., Schertzer, D., and Pecknold, S. 1996. Multifractal analysis and modeling of rainfall and river flows and scaling, causal transfer functions. *J. Geophys. Res.*, **101**, 26427–26440.

Thomson, W. 1887. On the propagation of laminar motion through a turbulently moving inviscid liquid. *Philos. Mag.*, 342–353.

Toschi, F., and Bodenschatz, E. 2009. Lagrangian properties of particles in turbulence. *Annu. Rev. Fluid Mech.*, **41**, 375–404.

Toschi, F., Biferale, L., Boffetta, G., Celani, A., Devenish, B. J., and Lanotte, A. 2005. Acceleration and vortex filaments in turbulence. *J. Turbul.*, **6**(6), 15.

Tran, T., Chakraborty, P., Guttenberg, N., Prescott, A., Kellay, H., Goldburg, W., Goldenfeld, N., and Gioia, G. 2010. Macroscopic effects of the spectral structure in turbulent flows. *Nature Phys.*, **6**(6), 438–441.

Tsang, Y. K., Ott, E., Antonsen Jr, T. M., and Guzdar, P. N. 2005. Intermittency in two-dimensional turbulence with drag. *Phys. Rev. E*, **71**(6), 066313.

Tsinober, A. 2009. *An informal conceptual introduction to turbulence*. Dordrecht: Springer Verlag.

Turiel, A., Nieves, V., García-Ladona, E., Font, J., Rio, M.-H., and Larnicol, G. 2009. The multifractal structure of satellite sea surface temperature maps can be used to obtain global maps of streamlines. *Ocean Science*, **5**(4), 447–460.

Uchaikin, V. V., and Zolotarev, V. M. 1999. *Chance and stability: stable distributions and their applications*. Walter de Gruyter.

Van Heijst, G. J. F., and Clercx, H. J. H. 2009. Laboratory modeling of geophysical vortices. *Annu. Rev. Fluid Mech.*, **41**, 143–164.

Veltcheva, A. D., and Soares, C. G. 2004. Identification of the components of wave spectra by the Hilbert Huang transform method. *Appl. Ocean Res.*, **26**(1-2), 1–12.

Vicsek, T., and Barabasi, A.-L. 1991. Multi-affine model for the velocity distribution in fully turbulent flows. *J. Phys. A*, **24**(15), L845.

Warhaft, Z. 2000. Passive scalars in turbulent flows. *Annu. Rev. Fluid Mech.*, **32**(1), 203–240.

Weisse, Ralf. 2010. *Marine climate and climate change: storms, wind waves and storm surges*. Springer Science & Business Media.

Wendt, H., Abry, P., and Jaffard, S. 2007. Bootstrap for Empirical Multifractal Analysis with Application to Hydrodynamic Turbulences. *IEEE Signal Processing Mag.*, **24**(4), 38–48.

Wilcox, D. C., et al. 1998. *Turbulence modeling for CFD*. Vol. 2. DCW industries La Canada, CA.

Wu, Z., and Huang, N. E. 2004. A study of the characteristics of white noise using the empirical mode decomposition method. *Proc. R. Soc. A*, **460**, 1597–1611.

Wu, Z., and Huang, N. E. 2010. On the filtering properties of the empirical mode decomposition. *Adv. Adapt. Data Anal.*, **2**(04), 397–414.

Wu, Z., Huang, N. E., Long, S. R., and Peng, C. K. 2007. On the trend, detrending, and variability of nonlinear and nonstationary time series. *PNAS*, **104**(38), 14889.

Xi, H. D., and Xia, K. Q. 2007. Cessations and reversals of the large-scale circulation in turbulent thermal convection. *Phys. Rev. E*, **75**(6), 066307.

Xi, H. D., Lam, S., and Xia, K. Q. 2004. From laminar plumes to organized flows: the onset of large-scale circulation in turbulent thermal convection. *J. Fluid Mech.*, **503**, 47–56.

Xia, H., Punzmann, H., Falkovich, G., and Shats, M. G. 2008. Turbulence-condensate interaction in two dimensions. *Phys. Rev. Lett.*, **101**(19), 194504.

Xia, H., Byrne, D., Falkovich, G., and Shats, M. 2011. Upscale energy transfer in thick turbulent fluid layers. *Nature Phys.*, **7**(4), 321–324.

Xu, H. T., Bourgoin, M., Ouellette, N. T., and Bodenschatz, E. 2006a. High order Lagrangian velocity statistics in turbulence. *Phys. Rev. Lett.*, **96**(2), 024503.

Xu, H. T., Ouellette, N. T., and Bodenschatz, E. 2006b. Multifractal dimension of lagrangian turbulence. *Phys. Rev. Lett.*, **96**(11), 114503.

Yaglom, A. M. 1957. Some classes of random fields in n-dimensional space, related to stationary random processes. *Theor. Probab. Appl+*, **2**, 273–320.

Yaglom, A. M. 1966. The influence on the fluctuation in energy dissipation on the shape of turbulent characteristics in the inertial interval. *Soviet Physics Dokladi*, **2**, 26–30.

Yeung, P. K. 2002. Lagrangian investigations of turbulence. *Annu. Rev. Fluid Mech.*, **34**(1), 115–142.

Zhang, Q., Xu, C., Chen, Y. D., and Yu, Z. 2008. Multifractal detrended fluctuation analysis of streamflow series of the Yangtze River basin, China. *Hydrol. Process.*, **22**, 4997–5003.

Zhao, X., and He, G.-W. 2009. Space-time correlations of fluctuating velocities in turbulent shear flows. *Phys. Rev. E*, **79**, 046316.

Zhou, Q., and Xia, K.-Q. 2008. Comparative experimental study of local mixing of active and passive scalars in turbulent thermal convection. *Phys. Rev. E*, **77**, 056312.

Zhou, Q., and Xia, K.-Q. 2011. Disentangle plume-induced anisotropy in the velocity field in buoyancy-driven turbulence. *J. Fluid Mech.*, **684**, 192–203.

Zhou, Q., Sun, C., and Xia, K.-Q. 2007. Morphological evolution of thermal plumes in turbulent Rayleigh-Bénard convection. *Phys. Rev. Lett.*, **98**, 074501.

Zhou, Q., Xi, H. D., Zhou, S. Q., Sun, C., and Xia, K. Q. 2009. Oscillations of the large-scale circulation in turbulent Rayleigh–Bénard convection: the sloshing mode and its relationship with the torsional mode. *J. Fluid Mech.*, **630**, 367–390.

Zhou, Q., Li, C. M., Lu, Z. M., and Liu, Y. L. 2011. Experimental investigation of longitudinal space-time correlations of the velocity field in turbulent Rayleigh-Bénard convection. *J. Fluid Mech.*, **683**, 94–111.

Zhou, S.-Q., and Xia, K.-Q. 2001. Scaling properties of the temperature field in convective turbulence. *Phys. Rev. Lett.*, **87**, 064501.

Zhou, S.-Q., and Xia, K.-Q. 2002. Plume statistics in thermal turbulence: mixing of an active scalar. *Phys. Rev. Lett.*, **89**(18), 184502.

Zolotarev, V. M. 1986. *One-dimensional stable distributions*. Vol. 65. American Mathematical Society.

Zongo, S. B, and Schmitt, F. G. 2011. Scaling properties of pH fluctuations in coastal waters of the English Channel: pH as a turbulent active scalars. *Nonlinear Proc. Geoph.*, **18**, 829–839.

Zongo, S. B., Schmitt, F. G., and Lefebvre, A. 2011. Observations biogéochimiques des eaux côtières à Boulogne-sur-mer à haute fréquence: les measures automatiques de la bouée MAREL,. 253–266.

Zybin, K. P., Sirota, V. A., Ilyin, A. S., and Gurevich, A. V. 2008. Lagrangian statistical theory of fully developed hydrodynamical turbulence. *Phys. Rev. Lett.*, **100**(17), 174504.

Index

algorithm, 8, 67, 89, 90, 92, 95, 97, 106, 110, 117, 126, 141, 149, 184
annual cycle, 3, 4, 10, 56, 60, 61, 122, 171, 172, 174
atmospheric, 14, 162, 169, 175, 179, 180
autocorrelation, 11, 22, 23, 63–67, 69, 70, 99, 106–109, 128, 147, 150, 164

bare, 27, 28

cascade, 4, 5, 10, 15–18, 21–24, 27–30, 32–34, 37, 39, 47, 48, 50, 51, 53, 59, 152–154, 156–160, 178
causal, 48
chaos, 24, 28
characteristic function, 22, 23, 26, 29, 30, 43, 47
cocumulative function, *see* cumulative function
codimension, 24, 25, 32, 34, 35
complex system, 3, 4, 6, 10, 100, 162
concave, 37, 46, 165, 183
cone, 48, 49
cumulative function, 59, 67, 74–76, 99, 100, 103, 104, 125, 138, 148, 156, 161

deterministic, 3, 4, 13, 14, 162, 165, 166, 173, 174, 183
detrended fluctuation analysis, 10, 11, 62, 74, 77, 112, 113, 115, 134, 160
detrended structure function, 56, 77, 78, 99, 110, 113, 130, 164
detrending, 76–78, 82, 83, 110, 130
dissipation, 5, 15, 17–21, 23, 27, 32, 33, 35, 36, 40, 106, 112, 136–138, 143, 146, 152
distribution, 1, 2, 19, 24, 26, 29–31, 42, 43, 45, 47, 70, 71, 83, 94, 95, 116, 122, 132, 140, 141, 153, 157
divergence of moments, 28, 36, 44
dressed, 27, 28, 30, 33, 37

eddy, 15, 18, 21
Empirical mode decomposition, 11, 41, 77, 89, 90, 92, 93, 97, 100, 116, 117, 131, 140, 157, 169
enstrophy, 59, 106, 112, 152–154, 156–159
ESS, *see* extended-self-similarity

Eulerian, 5, 37, 39, 40, 122, 160
extended-self-similarity, 106, 112, 122, 126, 130, 138, 139, 142, 156

finance, 50
fluctuation, 3, 4, 10, 15–21, 24, 27, 31, 32, 34, 35, 37, 39–41, 47, 57, 58, 72, 78, 82, 92, 100, 144, 151–154, 165, 169, 171, 176, 179, 180, 184
Fourier
    Fourier analysis, 6–10, 83, 85, 89, 90, 97, 100, 103, 113, 123, 128, 168, 171
    Fourier transform, 7, 64, 83, 85, 93, 104
fractal dimension, 24, 29
fractional Brownian motion, 33, 42, 44, 45, 50, 52, 119, 122, 129

gamma function, 58
gaussian, 31, 32, 42, 43, 47, 70, 116, 132, 140, 141, 153, 157, 158
geophysics, 3, 11
geoscience, 3, 10, 37, 56, 60, 61, 122

Hilbert
    Hilbert spectral analysis, 11, 62, 89, 93–95, 97, 133, 140, 158, 164, 170, 171, 174, 177, 182, 183
    Hilbert transform, 89, 93
    Hilbert-based method, 10, 56, 69, 94, 95, 97, 100, 121, 136, 142, 152, 159, 161
    Hilbert-huang transform, 10, 56, 89, 116
Hurst, 39, 50, 62, 67, 70–72, 75, 78, 79, 87, 88, 97, 178

iMF, *see* Intrinsic Mode Function
increment, 11, 32, 37–39, 41–43, 45, 50, 63, 64, 67, 70, 99, 106, 109, 151
inertial range, 4, 15–17, 23, 33, 39, 57, 60, 65, 103, 104, 106, 107, 110, 113, 117, 119, 125, 126, 128, 136–144
infinitely divisible, 30, 47
injection scale, 15, 22, 33

intermittency, *see* intermittent
intermittent, 1, 2, 5, 6, 10, 14, 15, 18, 20, 21, 23, 24, 27, 33, 37–40, 52, 55, 72, 97, 106, 107, 110–113, 115, 121–123, 126–137, 139, 142, 144, 150, 153, 154, 157, 159, 169, 174, 175, 177–179, 184
intrinsic mode function, 89–92, 93–98, 116, 132, 140, 157
isotropic, 10, 12, 15, 36, 56, 59, 72, 115, 126, 136, 138

kinetic energy, 14, 40, 78, 110
kolmogorov scale, 13, 17, 18, 33, 39

Lévy stable, 31, 43–45
Lagrangian, 2, 3, 5, 14, 39, 40, 59, 60, 62, 95, 97, 101, 122, 135–140, 142, 143, 152, 160
Legendre transform, 25, 26, 34, 35
log-poisson, 31, 34, 36, 159
log-stable, 31, 34
lognormal, 19, 20, 23, 26, 31, 33, 34, 47, 50, 53, 55, 106, 111–113, 115, 121, 122, 126, 127, 130, 134
long-memory, 45
long-range, 20, 23, 45, 48, 49

marine, 3, 36, 162, 163, 179
moment, *see* statistical moment
moment function, 25, 27–29, 32, 33, 35, 37, 41, 42, 44–46, 52, 55
multiaffine, 46, 50, 51
multifractal, 10, 11, 20, 22–26, 28, 31, 32, 40, 45–48, 50–53, 55, 62, 71, 74, 100, 123, 139, 140, 142, 143, 162, 175, 184
multifractality, 127, 130, 133, 157, 159, 160, 165
multiplicative, 20, 47, 50
  multiplicative cascade, 20, 22–24, 27, 28, 31, 32, 34, 39, 40, 47, 48, 50, 53
  multiplicative process, 32
multiscaling, 4, 10, 50, 56

Navier-stokes, 5, 13–16, 36, 154
nonlinearity, 2, 4, 8, 14
nonstationarity, 1, 2, 4, 179

ocean, 16–18, 152, 162
oceanic, 10, 14, 162, 165
oxygen, 1, 162, 171–174

passive scalar, 14, 32, 36, 37, 39, 40, 60, 76, 123, 127, 134, 135, 140, 144, 145, 153, 160, 162, 169, 171
Peclet, 36
power spectrum, 5, 9, 16, 18, 20, 33, 36, 38, 40, 41, 44, 57, 58, 61, 64, 67, 77, 78, 80, 82, 84, 101, 103, 116, 124, 130, 133, 136, 147, 152–154, 162, 163, 168, 175, 176, 181
power-law, 5, 20, 23, 34, 49, 56, 57, 59, 62, 63, 67, 68, 76, 78, 79, 87, 88, 94, 95, 97, 102, 105, 107, 109, 110, 113, 114, 117, 119, 121, 122, 124–126, 128–131, 133, 136, 139, 141, 142, 147–150, 152–154, 156–158, 163, 175, 176, 181
probability density function, 11, 19, 20, 69–72, 94, 95, 99, 110, 117, 128, 129, 132, 137, 140, 151, 153, 154, 157, 167, 172

ramp-cliff, 60, 76, 123–128, 130, 131, 134, 135, 160
random
  random measure, 47, 48
  random walk, 11, 43, 46, 47, 50, 51, 55
Rayleigh-Bénard, 59, 72, 101, 122, 144, 160, 161
Refined similarity hypothesis, 19, 106, 112, 136

salinity, 36, 172, 173
sampling, 8, 28, 36, 101, 124, 136, 146, 170, 176, 180
scalar variance, 36, 37, 40
scale
  scale invariance, 75, 94, 156
  scale ratio, 13, 14, 16, 22, 24, 27, 30, 48, 50
scaling
  scaling exponent, 19, 25, 34, 36, 39, 40, 50, 52, 57, 62, 63, 67, 72, 75–79, 81, 85, 87, 88, 94, 95, 97, 102, 105–107, 109–113, 115, 117, 119, 121–123, 125–131, 133, 135, 136, 138–142, 147, 150, 151, 153–155, 157–159, 169, 174, 177, 178, 182
  scaling process, 41, 42, 57, 62, 64, 71, 77–79, 82, 87, 88, 95, 107
singularity spectrum, 24, 79, 106, 112, 113, 115, 127, 130, 131, 134, 135, 140, 150, 159, 169, 183
statistical moment, 14, 16, 19, 21, 22, 25–28, 31, 33–36, 38, 39, 41–44, 81, 88, 94, 122, 138, 140, 157, 158, 169, 174, 177, 182, 183
stochastic
  stochastic integral, 43, 44, 52, 53
  stochastic process, 10, 23, 32, 41, 42, 46, 47, 49, 52, 119
strange attractor, 24, 25
structure function, 10, 11, 15, 19, 32–34, 36–38, 40, 41, 56, 58–60, 62, 67, 69, 72, 74–77, 79, 82, 83, 86, 87, 94, 95, 99, 100, 102–105, 107, 109, 110, 112, 123, 125, 126, 128, 134, 136–139, 144, 145, 147–149, 151, 153, 155, 156, 159–161, 164, 176, 178, 182, 183
subordinated, 45, 50, 51

Taylor
  Taylor's hypothesis, 37–39, 57, 101
  Taylor's microscale, 5, 101, 124
time-frequency, 6, 7, 93, 100

viscosity, 4, 12, 15, 17, 135, 146, 152–154

Wiener, 42, 51, 52, 57, 64

Printed in the United States
by Baker & Taylor Publisher Services